Computational Social Sciences

Simulating the Past

This innovative series publishes current international research and instruction on the methodological and theoretical aspects of quantitative and computer simulations in socio-historical contexts. Books published in this series bring together theories, techniques and technologies from backgrounds such as history, ecology, archaeology, anthropology, geography and sociology, combining them with established methodologies from the computational and complex systems sciences in order to promote deeper understanding and collaboration within the study of past human behavior and history.

Situated at one of the most dynamic interfaces of the emerging field of digital humanities and hard sciences, this series covers existing and novel computational approaches in connection with the presentation of case studies from numerous regions of the world and all prehistoric or historic periods. Particular topics of interest include, but are not limited to: have applications of computational modeling in archaeology and history, social organization and change, as well as socio-cultural transmission and evolution.

The audience for all books within this series consists of researchers, professionals and academics.

More information about this series at http://www.springer.com/series/15553

Mehdi Saqalli • Marc Vander Linden
Editors

Integrating Qualitative and Social Science Factors in Archaeological Modelling

Editors
Mehdi Saqalli
UMR CNRS 5602 GEODE
Géographie de l'Environnement
Maison de la Recherche
Université Toulouse 2 Jean Jaurès
Toulouse, France

Marc Vander Linden
Department of Archaeology
University of Cambridge
CB2 3Dz
Cambridge, UK

ISSN 2509-9574 ISSN 2509-9582 (electronic)
Computational Social Sciences
ISSN 2662-3145 ISSN 2662-3153
Simulating the Past
ISBN 978-3-030-12722-0 ISBN 978-3-030-12723-7 (eBook)
https://doi.org/10.1007/978-3-030-12723-7

© Springer Nature Switzerland AG 2019
This work is subject to copyright. All rights are reserved by the Publisher, whether the whole or part of the material is concerned, specifically the rights of translation, reprinting, reuse of illustrations, recitation, broadcasting, reproduction on microfilms or in any other physical way, and transmission or information storage and retrieval, electronic adaptation, computer software, or by similar or dissimilar methodology now known or hereafter developed.
The use of general descriptive names, registered names, trademarks, service marks, etc. in this publication does not imply, even in the absence of a specific statement, that such names are exempt from the relevant protective laws and regulations and therefore free for general use.
The publisher, the authors, and the editors are safe to assume that the advice and information in this book are believed to be true and accurate at the date of publication. Neither the publisher nor the authors or the editors give a warranty, express or implied, with respect to the material contained herein or for any errors or omissions that may have been made. The publisher remains neutral with regard to jurisdictional claims in published maps and institutional affiliations.

This Springer imprint is published by the registered company Springer Nature Switzerland AG
The registered company address is: Gewerbestrasse 11, 6330 Cham, Switzerland

Contents

Introducing Qualitative and Social Science Factors in Archaeological Modelling: Necessity and Relevance.................. 1
Marc Vander Linden and Mehdi Saqalli

***O Tempora O Mores*: Building an Epistemological Procedure for Modeling the Socio-anthropological Factors of Rural Neolithic Socio-ecological Systems: Stakes, Choices, Hypotheses, and Constraints**.. 15
Mehdi Saqalli, Melio Saenz, Mahamadou Belem, Laurent Lespez, and Samuel Thiriot

From Culture Difference to a Measure of Ethnogenesis: The Limits of Archaeological Inquiry............................. 55
Juan A. Barceló, Florencia Del Castillo, Laura Mameli, Franceso J. Miguel, and Xavier Vilà

Modeling Niche Construction in Neolithic Europe.................. 91
R. Alexander Bentley and Michael J. O'Brien

What Can a Multi-agent System Tell Us About the Bantu Expansion 3,000 Years Ago?.................................. 109
Florent Le Néchet, Christophe Coupé, Hélène Mathian, and Lena Sanders

From Past to Present: The Deep History of Kinship.............. 137
Dwight W. Read

Modeling the Relational Structure of Ancient Societies through the *Chaîne opératoire*: The Late Chalcolithic Societies of the Southern Levant as a Case Study......................... 163
Valentine Roux

Ethnoarchaeology-Based Modelling to Investigate Economic Transformations and Land-Use Change in the Alpine Uplands......... 185
Francesco Carrer, Graeme Sarson, Andrew Baggaley, Anvar Shukurov, and Diego E. Angelucci

Trowels, Processors and Misunderstandings: Concluding Thoughts 217
Mehdi Saqalli and Marc Vander Linden

Index... 225

Contributors

Diego E. Angelucci Dipartimento di Lettere e Filosofia, Università degli Studi di Trento, Trento, Italy

Andrew Baggaley School of Mathematics Statistics and Physics, Newcastle University, Newcastle upon Tyne, UK

Juan A. Barceló Universitat Autonoma de Barcelona, Barcelona, Spain

Mahamadou Belem Université Nazi Boni, École Supérieure d'informatique, Bobo Dioulasso, Burkina Faso

R. Alexander Bentley Department of Anthropology, University of Tennessee, Knoxville, TN, USA

Francesco Carrer McCord Centre for Landscape, School of History Classics and Archaeology, Newcastle University, Newcastle upon Tyne, UK

Christophe Coupé Department of Linguistics, The University of Hong Kong, Hong Kong, China

Florencia Del Castillo CONICET- Centro Nacional Patagónico (CCT-CENPAT-Argentina), Puerto Madryn, Argentina

Florent Le Néchet Université Paris-Est, Laboratoire Ville Mobilité Transport UMR-T 9403, UPEM, Marne-la-Vallée, France

Laurent Lespez LGP-UMR CNRS 8591, Université de Paris Est- Créteil & Université Paris 1, Paris, France

Laura Mameli Universitat Autònoma de Barcelona, Barcelona, Spain

Hélène Mathian UMR 5600 Environnement Ville Société, CNRS/Université de Lyon, Lyon, France

Franceso J. Miguel Universitat Autònoma de Barcelona, Barcelona, Spain

Michael J. O'Brien Department of Arts and Humanities, Texas A&M University–San Antonio, San Antonio, TX, USA

Dwight W. Read Department of Anthropology, University of California, Los Angeles, Los Angeles, CA, USA

Valentine Roux UMR CNRS, UMR 7055, University Paris-Nanterre, Nanterre, France

Melio Saenz Programa para el Manejo del Agua y del Suelo (PROMAS), University of Cuenca, Cuenca, Ecuador

Lena Sanders UMR 8504 Géographie-Cités, CNRS-Université Paris I-Université Paris VII, Paris, France

Mehdi Saqalli UMR CNRS 5602 GEODE Géographie de l'Environnement, Maison de la Recherche, Université Toulouse 2 Jean Jaurès, Toulouse, France

Graeme Sarson School of Mathematics Statistics and Physics, Newcastle University, Newcastle upon Tyne, UK

Anvar Shukurov School of Mathematics Statistics and Physics, Newcastle University, Newcastle upon Tyne, UK

Samuel Thiriot EDF Lab Paris-Saclay, Palaiseau, France

Marc Vander Linden Department of Archaeology, University of Cambridge, CB2 3DZ, Cambridge, UK

Xavier Vilà Universitat Autònoma de Barcelona, Barcelona, Spain

Introducing Qualitative and Social Science Factors in Archaeological Modelling: Necessity and Relevance

Marc Vander Linden and Mehdi Saqalli

Introduction

The early use of computational modelling and simulations in archaeology can be traced back to the 1960s and 1970s. These can be linked to both new computerizing technologies and concepts, but also to expanding processual paradigms which were dramatically instrumental in reshaping the archaeology discipline as we largely know and practise it still nowadays (Doran and Hodson 1975).

Whilst the acknowledgement of the role of computational modelling has experienced major fluctuations throughout the 1980s and 1990s (Gilbert and Troitzsch, 2005; Gilbert, 2006), parallel to major dissensions within archaeological theory, the use and relevance of computational models in archaeology has exponentially grown over the past couple of decades (for an extensive historical review, see Lake 2014, and for new explorations, see Barceló and del Castillo 2016). The reasons for this resurgence are manifold, and the growing availability of increasingly cheaper and powerful computing hardware must not be underestimated in this process (Lake 2014; Grosman 2016).

From a theoretical and methodological point of view, modelling approaches allow to articulate in a formal way numerous and various factors (Shennan and Steele 2005; Kohler and Gumerman 2000; Kohler and van der Leeuw 2007) and, especially in the case of agent-based models, to explore their complex interactions in a nondeterministic way (Cegielski and Rogers 2016). In this perspective, It is

---

M. Vander Linden
Department of Archaeology, University of Cambridge, CB2 3DZ, Cambridge, UK

M. Saqalli (✉)
UMR CNRS 5602 GEODE Géographie de l'Environnement, Maison de la Recherche, Université Toulouse 2 Jean Jaurès, Toulouse, France
e-mail: mehdi.saqalli@univ-tlse2.fr

noteworthy that computational models do not differ from so-called literary models, widely used by "traditional" archaeologists, in the sense of "verbal descriptions of a proposed set of causal relationships" (Steele and Shennan 2009: 108).

So, in regard of this apparent proximity, why does computational modelling remain a relatively niche and specialized activity within archaeology? For instance, Lake was able to count only 70 simulation studies having been published between 2001 and 2014, not counting purely conceptual and duplicates of the same model (Lake 2014). One may use as a counterexample the progressive intrication between GIS and geography studies to form one sole discipline, inducing the subsequent equivalent dynamic for geographically related disciplines such as geology, geomorphology or pedology. The same impregnation occurs between biology and related disciplines in one hand and computerized statistics on the other hand.

Of course, because modelling is also a question of practice for apprehending not-so-obvious concepts and acquiring related skills, many archaeologists face difficulties in terms of computer use and coding. Whilst this is arguably often the case, successful alternative strategies exist for formalizing the related interdisciplinarity (Etienne et al. 2011), and in numerous instances, computational modelling in archaeology is the outcome of productive, if complex, interdisciplinary projects. We rather hypothesize that more fundamental issues and misunderstandings about the modelling process, deeply rooted in theoretical and epistemological assumptions, explain the reluctant engagement of many archaeologists towards computational models.

As any in-depth assessment of either the full breadth of archaeological theory or archaeological computational modelling lies beyond the remits of the present text, the following introduction focuses on what seems to us to be two related issues, namely, perception of simplicity vs. simplicity as requirement and question of adequation of data and simulations.

A further remark is needed. As part of this volume and introduction, our attention is mostly devoted a particular time of computational models, namely, agent-based or multi-agent models (Cegielski and Rogers 2016). It must be stressed that this particular category of models is not, by a long margin, the only one used in archaeology as, for instance, reaction-diffusion models have been very popular when discussing the spread of populations and/or next technologies (Steele 2009 for a recent synthesis). The issues discussed here actually apply to all forms of modelling, though they are perhaps more acute when considered from the point of view of agent-based models.

Model Simplicity and Data

For many archaeologists, especially those less familiar with the approach, it seems that the simplicity of models is the most detrimental element, leading to a perception that models are mechanistic, inherently flawed with self-fulfilling prophecies and incapable to encompass in anyway the complexity of both the past and data.

Pointillism vs. impressionism: There is here a simple but meaningful difference about the definition of the search of truth: as a caricature, one may describe the field archaeologist as a pointillist approach, with a series of field dots, each one providing limited but significant information, eventually leading to a meaningful interpretation through accumulation of a sufficient number of dots. On the other hand, modellers would have an impressionist approach, where an initially gross, blurred but global image would improve along a more detailed process of local dynamics affecting the whole system. Therefore, one side sees models as always false and blurred, whilst the other side sees dots as – nearly – pointless.

Mimicking and not reproducing: Whilst several archaeologists consider simplification as a threat to the acknowledgement of the inherent complexity of human life, on the contrary modellers see it as a necessary step to get an insight into the very same complexity. In both cases thus, complexity is never challenged, although the ways to assess it profoundly differ. Computational models are, first and foremost, formal thought experiments: their mathematical architecture requires explicit formalism to translate hypotheses and suggested causalities into rules and code, but otherwise cannot, as already stated, have any claim to being intellectually better than literary models. In this sense, and if only for practical reasons, simplification is a necessary step. At least in modellers' mind, models do not aim at reproducing reality but rather, at best, at encapsulating some properties of reality.

"Kiss" vs. "Kids": The extent and nature of these properties are actually debated within the modelling community itself: some adopt a highly abstract stance and tend to consider models as pure theoretical, thought experiments, an approach often leading to simple models aiming at exploring one given hypothesis and/or process. On the contrary, other modellers consider that models must incorporate a higher level of complexity, leading to the addition of numerous parameters and rules. This somewhat binary dichotomy is often referred to as an opposition between so-called KISS (Keep It Simple Stupid) vs. KIDS (Keep It Descriptive Stupid) models (Edmonds and Moss 2005; Cegielski and Rogers 2016), and contributions to this volume stand on different sections of this spectrum, including the present authors (compare for instance Saqalli and Baum, 2016 with Drost and Vander Linden 2018).

Post hoc *ergo propter hoc:* This debate on the respective merits of simple vs. complex models is not simply to show that modellers are aware of how complex their object of study is. One may suggest repositioning this debate within the epistemological dialogue regarding the respective positioning of causalities and correlations into a scientific demarche: causalities link parameters through a dynamic, initially qualitatively defined process. But because causalities cannot be proven, they are hypothesized. Testing these hypotheses implies refutation using correlation significance and, therefore, the use of quantitative data.

Quantitative data are an imperative step for replicable science: they define the weight and values of the different parameters which, if non-characterized, would produce fuzzy and useless results. However, defining which parameters are required for archaeology turns out to be an unexpectedly extremely difficult task. For instance, several simulations presented here are spatially explicit and therefore include extensive discussions upon the selection and identity of spatial parameters under consideration, including presence/absence of specific features such as rivers, ecological niches or land use categories. As many authors here explain, such debates do occur between proponents of various disciplines and are, intrinsically, qualitative. Yet, it remains obvious that any simulation space, however complex it is, will never qualify as a proper ecology or as a landscape lived in and experienced by humans. This discrepancy may appear as a shortcoming of modelling, but is not detrimental since the model never aims at being real.

The second source of misunderstanding between modellers and archaeologists lies in their respective attitude towards data. Indeed, from a modelling perspective, the development of models does not merely rise, as sometimes may be thought by non-practitioners, from the sole wish to offer a formal description of past, but as a response to old archaeological problem of the generally low resolution and patchy character of the archaeological record. As stressed in several chapters (Saqalli et al., this volume; O'Brien and Bentley, Chap. 4, this volume; Carrer et al., Chap. 8, this volume), archaeological data are by definition flawed, and, following David Clarke's well-known definition, archaeology is the "discipline with the theory and practice for the recovery of unobservable hominid behavior patterns from indirect traces in bad samples" (Clarke 1973: 17). Whilst archaeologists have devised countless tools to account for some of these biases (e.g. taphonomy), eventually it remains impossible to assess how much exactly was lost from the past. The modellers' response to this conundrum is to perform simulations, meaning the creation of "fake" data. The point of the procedure is not to obtain a simulacrum of the past or, worse, the pretense of an alternative past but rather a continuous environment generated under known conditions. This decision in turn raises further problems as to how to compare such unbiased distribution with patchy archaeological data (see below).

Socio-Anthropology as an Unavoidable Science

The development of socioecological models, for both past and present situations, relies upon the selection of relevant variables. If the translation from quantitative data to components' behaviour rules is relatively easy, the translation from qualitative traits is yet to be settled. The latter is generally achieved by calculating mean or median values and variances from various existing datasets, which are then used to set parameters and/or rules in the models. The criteria applied in choosing relevant variables are often less explicit. This is well exemplified in models where environmental factors are often given primacy. Yet numerous studies indicate that anthropological factors (e.g. family organization, inheritance rules, distribution of power among lineages or families) can drastically impact upon the environment, sometimes in a counter-intuitive way (see, for instance, Rouchier and Requier-Desjardins 2000; Polhill et al. 2010; Saqalli et al. 2010). Therefore, we raise the more general question of the inclusion and consideration of qualitative variables into computational models and, beyond, of the scientific validation of such models. This proposed volume will explore these questions, including the corresponding theoretical and epistemological challenges.

Thanks to inference from better-known situations coming, for example, from the present time, where one may directly get access to people's beliefs, intentions and rationalities through a mix of socio-anthropological field methodologies (interviews, questionnaires, participatory observations, etc.), we are, mostly, aware of the importance of related factors not only in terms of conceptual weight but more prosaically in terms of evolution of the model. For instance, a society with a one-heir inheritance system in a context of land scarcity or with gender differentiation induces strong and self-amplifying social stratification without further dynamics. But how strong is this influence? Could these factors in general be not crucial enough for NOT being considered in socioecological models? Thanks to a wide array of paleo-environmental datasets (e.g. dendrochronology, carpology, palynology, anthracology and other data collecting methods on climate and vegetation), access to these elements are far easier. Suggesting causalities between climatic and/or environmental events and archaeological dynamics thus becomes very tempting, but does it pass the test of refutation, such as equivalent environmental events, without visible impacts on archaeological records? We assume that it cannot be. As a simple illustration, one may think of important historical events in well-archived societies with no possible connection to environmental dynamics. More globally, let us consider, for instance, rural societies where demographic density is low and where access to any asset (food, water, shelter, etc.) depends mainly on manpower. Access to manpower in a family is defined by marriages, demography, family organization, extra-family organization and rules, differential access to production activities, etc. In short, one's living and even surviving conditions depend on social rules within the related culture. Finally, one may find difficult to not acknowledge the importance of social networks (e.g. Roux, Chap. 7, this volume; see also Lemieux 1976; Collier et al. 2009; Amblard et al. 2010; Filho et al. 2011; Neumann et al. 2011; Gabbriellini 2014; Neumann and Lotzmann 2016; Thiriot 2018).

Once we assume such a position, how can we formalize such non-environmental factors relying on socio-anthropological and political factors? How can we assess their integration epistemologically, scientifically and practically? The following chapters offer numerous proposals for building such an integrative corpus of paradigms and methods.

A Brief Overview of Individual Chapters

The seven chapters in this volume all explore in different ways these many questions. Although a certain emphasis is discernible upon modelling of ecological-economic systems (Saqalli et al. this volume; Barceló et al., Chap. 3 this volume; Le Néchet et al., Chap. 5, this volume; Carrer et al., Chap. 8, this volume), the following chapters explore a wider range including niche construction theory (Bentley and

O'Brien, Chap. 4, this volume), kinship (read this volume) and cultural transmission (Roux, Chap. 7, this volume). A majority of chapters include several authors from varied background, reflecting the aforementioned strong interdisciplinary component of archaeological modelling, though it must be noted that single- and dual-authored chapters all have stem from longer collaborative research projects (e.g. Roux et al. 2018).

The opening chapter by Saqalli and colleagues recalls several points made earlier in this introduction, with further emphasis regarding the status of modelling within archaeology. For instance, they ask whether computational modelling should be considered as an integral part of the discipline or rather as a highly specialized activity "subcontracted" to external practitioners. Beyond the obvious relations of power both alternatives imply, they stress the centrality of interdisciplinarity in modelling, not simply because it relies upon different skill sets distributed among various fields but fundamentally because the research questions being addressed, especially so in rural socioecological systems, involve several disciplines and categories of evidence. Such interdisciplinary dialogue relies upon trust and humility between partners (see also Vander Linden 2017), as well as sacrifice on each partner's part. As Saqalli and colleagues put it, the need for sacrificing a portion of the complexity of each object of inquiry is imperative to assure interdisciplinary dialogue and to achieve the ultimate simplification requested by the very act of modelling. Although simplification is thus a necessary step, the approach adopted by Saqalli and colleagues to account for early European Neolithic farming systems (see also Saqalli et al. 2014) clearly falls towards the KIDS end of the spectrum, complexity being considered as necessary to capture the variability and adaptability of rural socioecological systems. Their chapter provides a series of practical recommendations for the elaboration of such models, including the choice of appropriate analogues for identifying and setting up parameters and the implementation of the rules governing the behaviour of the model. One of their key points is the dynamic nature of modelling. Indeed, because of their architecture, computational models allow for continuous modification, addition or suppression of parameters. Yet, this complexity comes at a prize, as the multiplicity of parameters renders difficult the evaluation of their confounding effects upon the behaviour of the simulations. Possible solutions include hierarchization of the parameters, required to avoid the pitfalls of determinism, the combination of factors belonging to different levels and sensitivity analysis. Alternative pathways include more parsimonious approaches based upon incremental additions of parameters, a process admittedly easier to undertake with KISS models (e.g. Drost and Vander Linden 2018).

Although focusing on an altogether radically different topic – ethnogenesis – Barceló and colleagues' contribution follows closely similar epistemological and methodological lines to the ones exposed by Saqalli and co-workers. Not only do they illustrate that modelling can address deeply humanistic questions such as identity, and thus need not being limited to economic and environmental issues (see also Kovačević et al. 2015), their chapter also discusses issues related to the way archaeologists routinely describe and transform their data and how modelling can contribute to such continuous methodological renewal. Their extensive and complex review of the literature on ethnicity and identity elegantly illustrates that the apparent simplicity of their model does not stem from unawareness of the issues at stake but rather of a conscious, deliberate methodological and epistemological reflexion. In this sense, their simulation approach expresses the close links between ethnicity, knowledge transmission and thus historicity. Arguably, reducing ethnicity to a single measure of similarity may seem reductionist, but it provides an effective way to conceptualize otherwise implicit interpretative themes, as well as a quantifiable measure inferred from numerous archaeological data, following David Clarke's famous polythetic definition of archaeological assemblages (Clarke 1968). This explicit reference to Clarke's work provides one of the many testimonies to his long-lasting influence upon the modelling community (e.g. recently Lycett and Shennan 2018). Barceló and colleagues' model also exemplifies how much ethnicity can only be studied in an effective way as the combination of observable practices, seen as a vast array of factors, all articulated in their model design (e.g. food acquisition and exchange, human mobility, varying carrying capacity of the environment). Whilst Barceló and colleagues are dedicated to analyse both simulation results and archaeological data using the same techniques (i.e. similarity indices), this methodological decision leads them to face the aforementioned problem of having to deal with very dissimilar datasets, with models full of data of the one hand and archaeological data "full of holes" on the other hand, notwithstanding the fact that several variables remain simply out of reach of the resolution provided by archaeological data. This is a recurring crucial issue and one which both provides the justification of computational models, but which can also be detrimental to their widespread application (beyond being "mere" thought experiments). There is probably no single answer on how to resolve this tension between "ideal" simulation data and "biased" archaeological data, but the recent literature offers several avenues worth exploring, including sampling of the simulation results mimicking archaeological deposits (Kovačević et al. 2015), approximate Bayesian computation (Edinborough et al. 2015), whilst Barceló and colleagues consider the role of bootstrapping.

Bentley and O'Brien also address the fundamental issue of the adequation between archaeological data and the requirements of modelling and statistical approaches. Following upon previous work (O'Brien and Bentley 2015; see also Bentley et al. 2015), their contribution lies firmly in the tradition of evolutionary archaeology and focuses upon niche construction during the European Neolithic, especially the possible co-evolution of dairying practices and lactase persistence (see also Itan et al. 2009; Mathieson et al. 2015). The concept of niche construction provides a robust framework for the coherent analysis of concepts otherwise often studied separately, for instance, cultural transmission and food production systems, thus re-affirming the potential for computational modelling to offer efficient tools to discuss unexpected series of relationships and to bridge the compartmentalization of disciplines. Rather than adopting a simulation approach strictly speaking, Bentley and O'Brien explore the relevance of Granger causality, a linear modelling technique used to assess the incremental predictive value of time series. The advantage of such technique lies in the limited number of variables being considered (here only two), but the drawback is the extreme demands in terms of data required to fix the equations. Despite being arguably one of the richest archaeological records available across the globe, with extensive datasets covering radiocarbon dates, cemeteries and mortality profiles and settlement patterns, Bentley and O'Brien end up concluding that available data for European Neolithic periods do not match up the requirements of the matrix.

Le Néchet and colleagues' chapter also weaves together modelling and food production systems, in this case through the lens of the Bantu expansion and in particular the interaction between Bantu farmers and forest foragers. A combination of archaeological, linguistic and genetic data provides them with competing hypotheses regarding the direction and specifics of the migration of Bantu-speaking farming populations and especially the role of non-Bantu foraging communities in mediating this process. Informed by historical and ethnographic resources, they stress the symbiotic relationship between both communities and lifeways, focused upon two distinct co-existing ecological niches. In this respect, the African situation described here markedly differs from more European-centric case studies and simulations, which often consider foragers and farmers as either antagonistic entities or as stages in a linear evolutionary process ultimately leading to the dominance of the latter over the former (e.g. Aoki et al. 1996). Despite this fundamental difference, their model also lies in the continuity of a long tradition of computational models exploring demographic expansion, often based upon reaction-diffusion equations and occurring in an empty, unrealistic landscape (see Steele 2009, Vander Linden and Silva 2018; noticeable exceptions include Davison et al. 2006, Fort et al. 2012). In contrast to such equations, Le Néchet and colleagues rather use a multi-agent system, not simply for the complexity inherent to such computational tools but because these are designed to explore emergent properties and thus lack any central control mechanism, echoing expectations regarding the historical situation of the Bantu spread. Interestingly, Le Néchet and colleagues explicitly wish to position their H.U.M.E. model (HUman Migration and Environment) in the middle of the KISS/KIDS dichotomy through the incorporation of both generic and specific traits. As a result, their model incorporates parameters related to food resources and productivity, group fission, technological innovation and group interaction (including competition and imitation). Rather than never-endingly debating upon what constitutes the minimum requirements for a model to be complex or realistic enough, there is, as Le Néchet and colleagues explain, more to be gained by using computational model as the basis for dynamic interdisciplinary dialogue (see also Saqalli et al. this volume; Carrer et al., Chap.8, this volume). Le Néchet and colleagues, for instance, depict how earlier versions of their model, and the underlying decisions structuring them, are questioned and transformed through interactions between various participants, eventually leading to the inclusion of several new variables (e.g. modelling two populations using same parameters but set with two distinct sets of values, spatially explicit environment considering distinct ecologies). Overall, they offer the description of an entire methodological protocol, designed to incorporate differences traditions of conceptualizing and analysing data.

Read's chapter may, at first sight, seem to stand out in this volume, as it does provide neither computational model nor any computer code nor any simulation. Elsewhere, though, he has provided a detailed demographic simulation of hunter-gatherer societies (Read 1998; see also Read and LeBlanc 2003). In his simulation, he shows that hunter-gatherer societies will have a stabilized population size substantially below carrying capacity (with carrying capacity based on the limitations of the resource procurement systems of hunter-gatherer groups) in regions with a low density of resources and a stabilized population size close to carrying capacity for groups in regions with a high density of resources. This leads to the prediction that intergroup violence is more likely for hunter-gatherer groups in regions with a high resource density. This prediction has been substantiated for hunter-gatherer groups in Australia (Read 2009). In this simulation, Read also showed that if marriages among the! Kung San are consistent with their cultural kinship rules regarding proper and improper marriages, then there would be de facto residence group marriage exogamy even though there is no cultural rule requiring that marriages be exogamous with respect to one's residence group. This shows the value of simulations for working out the behavioural consequences of cultural rules, thereby adding to the ethnographic record. With regard to the kinship terminology systems that Read discusses in this chapter, he has also developed (in conjunction with Michael D. Fischer at Kent University, UK; see Read et al. 2013) an extensive computer model, called Kinship Algebraic Expert System (KAES), for implementing the algebraic analysis of kinship terminology systems that he discusses in his chapter (Read 2006). The computer model makes it possible for the algebraic analysis of a kinship terminology to be carried out even without a background in the formalism of abstract algebras. In this chapter, his contribution demonstrates how mathematically driven simplification, implemented in the KAES computer programme, offers in-depth insights into such a seemingly complex and fundamental human process as kinship. Read shows how the specificity of human kinship systems rests upon their computational properties by introducing a fundamental qualitative distinction between interaction systems based upon individuals on the one hand and relation-based social systems on the other hand. Read's argument rests upon the characterization and description of the evolutionary sequence from genealogical connections (e.g. mother and father relationships) to the symbolic system of kinship relations.

Roux's chapter explores divergences between existing archaeological approaches on cultural transmission and technological change, in particular the respective assumptions of computational models rooted in evolutionary thinking, analytical sociology and the long tradition of ethnoarchaeological research on technology, especially the French *chaîne opératoire* approach. Roux's criticisms of evolutionary model lie upon the measure of evolution processes independently of the social context into which they take place, as well as upon morphometric traits which, despite having the advantage of being easy to quantify, are known to be poor markers of transmission and prone to rapid, quasi-stochastic changes. By contrast, Roux's preference lies with technical traits, as extensive ethnoarchaeological research demonstrates that their acquisition by social agents is tightly associated with social learning. For instance, pottery forming techniques are often acquired over a long period of time, required to master the necessary motor skills (Roux and Corbetta 1989; Roux et al. 1995 for a non-pottery example), and are thus intrinsically more stable than other parts of the *chaîne opératoire*. In this sense, the characterization of the transmission of any technology becomes directly linked to the evaluation of the social structure within which learning occurs. Such consideration for both the content of what is being transmitted and the associated mechanisms of social interaction finds strong parallels with analytical sociology, a field with a long tradition of involvement with simulation approaches, especially network models. The approach suggested here however qualitatively differs from most archaeological applications of network analysis, based on the assumption that similarity, interaction and exchange follow a linear relationship. Given that such material similarities can be the outcome of distinct processes, Roux insists upon the need to select as appropriate and as robust as possible variables or proxies for the subject to be modelled and explained. In this perspective, the selection of relevant proxies is geared at highlighting the relational structure of societies and then using reference sociological models providing explanation of evolution processes to analogue relational structures of societies. Here, Roux's epistemological and methodological reflection is put to practice by using the Late Chalcolithic period (4500–3900 cal BC) in the Levant as case study, eventually demonstrating the leading role of long-lasting social network structures in shaping processes of technological innovation and transfer. Although the modelling component is less apparent in her chapter, more explicit considerations can be found in recent publications by her research group (e.g. Manzo et al. 2018).

The last contribution to this volume, by Carrer and collaborators, combines computational modelling, considered as an exploratory framework, ethnoarchaeology, seen as a robust source of analogues for setting up parameters of the models, and historical data on land use in the North Italian Alps, used to validate the results of the simulation. Their starting point lies in the driving role of both environmental and human factors in shaping land use patterns, in particular in mountainous landscapes. Contrasting the long history of human presence in mountains, only accessible through low-resolution archaeological and paleo-ecological data, with the comparatively short-term historical ecological knowledge, they advocate the use of computational modelling to further our knowledge of land use over the *longue durée*. As many other chapters, their methodology is intrinsically interdisciplinary, with a greater emphasis upon the essential role of ethnoarchaeologically documented analogues to set up parameters and calibrate simulations (see also Lancelotti et al. 2017). Here they offer two different models, reflecting two contrasted land use strategies (i.e. local subsistence relying upon cereal cultivation and cattle husbandry and intensive dairy-focused cattle husbandry). Both models are spatially explicit and incorporate different ecological niches, and corresponding human practices and carrying capacity, much in line with similar decisions found elsewhere in this volume (especially Saqalli et al. this volume, Barceló et al., Chap. 3, this volume, Le Néchet et al., Chap. 5, this volume). Although admittedly – and explicitly simple – both models perform rather well, providing reasonable fits with independent historical land use data.

Conclusion

Although the contributions assembled here cover a range of theoretical approaches and topics, they all share core concerns related to the tension arising between the need for simple models and the inclusion of complex, qualitatively demanding parameters or between "ideal" simulation and "patchy" archaeological data. Another recurrent theme, inherently linked to the first issue, lies in the development of spatially explicit models, especially when dealing with rural ecological systems. As previously said, there is no one-size-fits-all solution to be sought or found in these pages but, more prosaically, numerous methodological and practical solutions. It would however be unrealistic to put such stress upon the modelling community, including its producers and consumers, to come with such magical recipes: the same way archaeological sites require constant development of new and adaptation of long-established digging techniques, computational modelling remains a work in progress.

In this sense, we cannot insist more how much computational modelling provides an exciting venue for future archaeological research, especially given its formal requirements and its near-intrinsic interdisciplinary. Although we therefore extend Lake's enthusiasm and call for further development and popularization of modelling in archaeology (Lake 2014), at the same time we are not advocating for modelling to become a norm or standard. Firstly, the approach is not appropriate for all dimensions of archaeological practice. Secondly, and despite the fact that computational modelling is not locked in any brand of archaeology, calling for the normalization of modelling would not be in tune with the diversity of theoretical approaches which has always be the hallmark of the discipline.

Yet we are confident that the variety of topics and approaches exposed here will provide another supplementary step in clearing some of the existing misunderstandings between archaeologists and modelling community and help foster new work. As argued earlier, the conditions for the successful integration of qualitative factors in modelling extend well beyond the immediate needs of modellers, but have profound implications for archaeological reasoning, including our ability to excavate, analyse and share data.

References

Amblard, F., Geller, A., Neumann, M., Srbljinovic, A., & Wijermans, N. (2010). Analyzing social conflict via computational social simulation: A review of approaches. *NATO Science for Peace & Security Series, 75*, 141. https://doi.org/10.1111/j.1468-2486.2009.00910.x.

Aoki, K., Shida, M., & Shigesada, N. (1996). Travelling wave solutions for the spread of farmers into a region occupied by hunter-gatherers. *Theoretical Population Biology, 50*(1), 1–17.

Barceló J.A., & del Castillo F. (2016). Simulating prehistoric and ancient worlds, Springer Computational Social Sciences. Springer, USA.

Bentley, R. A., O'Brien, M. J., Manning, K., & Shennan, S. (2015). On the relevance of the European Neolithic. *Antiquity, 89*(347), 1203–1210.

Cegielski, W. H., & Rogers, J. D. (2016). Rethinking the role of agent-based modeling in archaeology. *Journal of Anthropological Archaeology, 41*, 283–298.
Clarke, D. L. (1968). *Analytical archaeology*. London: Methuen.
Clarke, D. L. (1973). Archaeology: The loss of innocence. *Antiquity, 47*, 6–18.
Collier, N., Boedhihartono, A. K., & Sayer, J. (2009). *Indigenous livelihoods and the global environment: Understanding relationships*. Presented at the 18th World IMACS / MODSIM Congress, pp. 2833–2839.
Davison, K., Dolukhanov, P., Sarson, G. R., & Shukurov, A. (2006). The role of waterways in the spread of the Neolithic. *Journal of Archaeological Science, 33*(5), 641–652.
Doran James Edward, Hodson Frank Roy. (1975). Mathematics and computers in archaeology. Harvard University Press.
Drost, C. J., & Vander Linden, M. (2018). Toy Story: Homophily, transmission and the use of simple simulation models for assessing variability in the archaeological record. *Journal of Archaeological Method and Theory*. https://doi.org/10.1007/s10816-018-9394-y.
Edinborough, K., Crema, E. R., Kerig, T., & Shennan, S. (2015). An ABC of lithic arrowheads: A case study from southeastern France. In C. Brink, S. Hydén, K. Jennbert, L. Larsson, & D. Olausson (Eds.), *Neolithic diversities perspectives from a conference in Lund, Sweden (Acta Archaeologica Lundensia)* (Vol. 65, pp. 213–223).
Edmonds, B., & Moss, S. (2005). From KISS to KIDS: An "anti-simplistic" modelling approach. *Lecture Notes in Artificial Intelligence, 34*, 130–144.
Etienne, M., DuToit, D., & Pollard, S. (2011). ARDI: A co-construction method for participatory modelling in natural resources management. *Ecology and Society, 16*, 44.
Filho, H.S.B., Neto, F.B., Fusco, W., (2011). *Migration and social networks — An explanatory multi-evolutionary agent-based model*. In: Intelligent Agent (IA), 2011 IEEE Symposium. Intelligent Agent (IA), 2011 IEEE Symposium on, pp. 1–7. https://doi.org/10.1109/IA.2011.5953616
Fort, J., Pujol, T., & Vander, L. M. (2012). Modelling the neolithic transition in the Near East and Europe. *American Antiquity, 77*(2), 203–219.
Gabbriellini, S. (2014). The evolution of online forums as communication networks: An agent-based model. *Revue Française de Sociologie, 55*, 805–826. https://doi.org/10.3917/rfs.554.0805.
Gilbert Nigel. (2006). Sciences sociales computationnelles: simulation sociale multi-agents. In: Modélisation et simulations multi-agents: application pour les sciences de l'Homme et de la Société, Amblard Frédéric, Phan Denis, 141–59. Paris, France.
Gilbert Nigel, Troitzsch Klaus G. (2005). Simulation for the Social Scientist. Open University Press, Glasgow, UK.
Grosman Leore. (2016). Reaching the Point of No Return: The Computational Revolution in Archaeology. Annual Review of Anthropology 45, 1: 129–45. https://doi.org/10.1146/annurev-anthro-102215-095946.
Itan, Y., Powell, A., Beaumont, M. A., Burger, J., & Thomas, M. G. (2009). The origins of lactase persistence in Europe. *Plos Computational Biology, 5*(8), e1000491. https://doi.org/10.1371/journal.pcbi.1000491.
Kohler, T. A., & Gumerman, G. J. (2000). *Dynamics in human and primate societies: Agent-based modeling of social and spatial processes*. Oxford, UK: Oxford University Press.
Kohler, T. A., & van der Leeuw, S. E. (2007). *The model-based archaeology of socionatural systems*. Oxford, UK: Oxbow Books Ltd.
Kovačević, M., Shennan, S., Vanhaeren, M., d'Errico, F., & Thomas, M. G. (2015). Simulating geographical variation in material culture: Were early modern humans in Europe ethnically structured? In *Learning strategies and cultural evolution during the paleolithic* (pp. 103–120). Tokyo: Springer.
Lake, M. W. (2014). Trends in archaeological simulation. *Journal of Archaeological Method and Theory., 21*, 258–287.
Lancelotti, C., Negre, J., Alcaina Mateos, J., & Carrer, F. (2017). Intra-site spatial analysis in ethnoarchaeology. *Environmental Archaeology, 22*(4). https://doi.org/10.1080/14614103.2017.1299908.

Lemieux, V. (1976). L'articulation des réseaux sociaux. *Recherches sociographiques, 17*, 247–260. https://doi.org/10.7202/055716ar.

Lycett, S. J., & Shennan, S. J. (2018). David Clarke's analytical archaeology at 50. *World Archaeology.* https://doi.org/10.1080/00438243.2018.1470561.

Manzo, G., Gabbriellini, S., Roux, V., & Nkirote M'Mbogori, F. (2018). Complex contagions and the diffusion of innovations: Evidence from a small-N study. *Journal of Archaeological Method and Theory.* https://doi.org/10.1007/s10816-018-9393-z.

Mathieson, I., Lazaridis, I., Rohland, N., Mallick, S., Patterson, N., Roodenberg, S. A., et al. (2015). Genome-wide patterns of selection in 230 ancient Eurasians. *Nature, 528*(7583), 499–503.

Neumann, M., Braun, A., Heinke, E. M., Saqalli, M., & Srbljinovic, A. (2011). Challenges in modelling social conflicts: Grappling with polysemy. *Journal of Artificial Societies & Social Simulations, 14,* 9.

Neumann, M., & Lotzmann, U. (2016). Simulation and interpretation: A research note on utilizing qualitative research in agent-based simulation. *Journal of Swarm Intelligence and Evolutionary, 5.* https://doi.org/10.4172/2090-4908.1000129.

O'Brien, M. J., & Bentley, R. A. (2015). The role of food storage in human niche construction: An example from Neolithic Europe. *Environmental Archaeology, 20*(4), 364–378.

Polhill, G. J., Sutherland, L. A., & Gotts, N. M. (2010). Using qualitative evidence to enhance an agent-based modelling system for studying land use change. *Journal of Artificial Societies & Social Simulations, 13,* 10.

Read, D. (1998). Kinship based demographic simulation of societal processes. *Journal of Artificial Societies and Social Simulation.* www.soc.surrey.ac.uk/JASSS/1/1/1.html

Read, D., & LeBlanc, S. (2003). Population growth, carrying capacity, and conflict. *Current Anthropology., 44*(1), 59–85.

Read, D. (2006). Kinship Algebra Expert System (KAES): A software implementation of a cultural theory. *Social Science Computer Review, 24*(1), 43–67.

Read, D. (2009). Agent-based and multi-agent simulations: Coming of age or in search of an identity? *Computational and Mathematical Organization Theory, 16,* 329–347.

Read, D., Fischer, M., & Leaf, M. (2013). What are kinship terminologies, and why do we care? A computational approach to analysing symbolic domains. *Social Science Computer Review, 31*(1), 16–44.

Rouchier, J., & Requier-Desjardins, M. (2000). La modélisation comme soutien à l'interdisciplinarité en recherche-développement. Une application au pastoralisme soudano-sahélien. *Nature, Sciences & Société, 8,* 61–67.

Roux, V., Bril, V., & Dietrich, G. (1995). Skills and learning difficulties involved in stone knapping: The case of stone-bead knapping in Khambat, India. *World Archaeology, 27,* 63–87.

Roux, V., & Corbetta, D. (1989). *The Potter's wheel. Craft specialization and technological competence.* Oxford: IBH Publishing.

Roux, V., Bril, B., & Karasik, A. (2018). Weak ties and expertise: Crossing technological boundaries. *Journal of Archaeological Method and Theory, 25,* 1024. https://doi.org/10.1007/s101816-018-9397-9.

Saqalli, M., Gérard, B., Bielders, C. L., & Defourny, P. (2010). Testing the impact of social forces on the evolution of Sahelian farming systems: A combined agent-based modeling and anthropological approach. *Ecological Modelling, 221,* 2714–2727. https://doi.org/10.1016/j.ecolmodel.2010.08.004.

Saqalli, M., Salavert, A., Bréhard, S., Bendrey, R., Vigne, J.-D., & Tresset, A. (2014). Revisiting and modelling the woodland farming system of the early Neolithic Linear Pottery Culture (LBK), 5600–4900 BC. *Vegetation History and Archaeobotany, 23*(S1), 37–50.

Saqalli Mehdi, Baum Tilman Georg. (2016). Pathways for scale and discipline reconciliation: Current socioecological modelling methodologies to explore and reconstitute human prehistoric dynamics. In: Simulating prehistoric and ancient worlds, Barceló Juan Antonio, del Castillo Florencia, 233-55. Springer Computational Social Sciences. Springer.

Shennan, S., & Steele, J. (2005). *The archaeology of human ancestry power, sex and tradition.* London: Routledge.

Steele, J. (2009). Human dispersals: Mathematical models and the archaeological record dispersal models and case studies: Fisher-Skellam-KPP. *Human Biology, 81*, 121–140.

Steele, J., & Shennan, S. (2009). Introduction: Demography and cultural macroevolution. *Human Biology, 81*, 105–119.

Thiriot, S. (2018). Word-of-mouth dynamics with information seeking: Information is not (only) epidemics. *Physica A: Statistical Mechanics and its Applications, 49*(2), 418–430. https://doi.org/10.1016/j.physa.2017.09.056.

Vander Linden, M. (2017). Reaction to a reactionary text. *Norwegian Archaeological Review, 50*(2), 127–129.

Vander Linden, M., & Silva, F. (2018). Comparing and modeling the spread of early farming across Europe. *PAGES News, 26*(1), 28–29.

O Tempora O Mores: Building an Epistemological Procedure for Modeling the Socio-anthropological Factors of Rural Neolithic Socio-ecological Systems: Stakes, Choices, Hypotheses, and Constraints

Mehdi Saqalli, Melio Saenz, Mahamadou Belem, Laurent Lespez, and Samuel Thiriot

Introduction

Since about two decades, researchers build models of past rural socio-ecological systems (RSES). These models are the result of the intersection of *archaeology*, which gathers and interprets remains of these past systems in order to understand better our past, and *modeling and simulation* which comes with its own concepts, methodologies, and practices. The irruption of computational modeling in a domain of social sciences is never straightforward. Indeed, it has to be discussed in order to explicit and bind the epistemological role of modeling for this field according to its peculiarities and traditional methodologies such as equivalent dynamics occurring in other disciplines such as geography (Lambin et al. 2000, 2001), economics

---

M. Saqalli (✉)
UMR CNRS 5602 GEODE Géographie de l'Environnement, Maison de la Recherche, Université Toulouse 2 Jean Jaurès, Toulouse, France
e-mail: mehdi.saqalli@univ-tlse2.fr

M. Saenz
Programa para el Manejo del Agua y del Suelo (PROMAS), University of Cuenca, Cuenca, Ecuador

M. Belem
Université Nazi Boni, École Supérieure d'informatique, Bobo Dioulasso, Burkina Faso

L. Lespez
LGP-UMR CNRS 8591, Université de Paris Est- Créteil & Université Paris 1, Paris, France
e-mail: laurent.lespez@lgp.cnrs.fr

S. Thiriot
EDF Lab Paris-Saclay, Palaiseau, France
e-mail: samuel.thiriot@edf.fr

(Livet et al. 2014), socio-environmental psychology (Ostrom 1988), social sciences in general (Epstein 1999; Gilbert and Troitzsch 2005), sociology (Epstein 2008), and political sciences (Cioffi-Revilla and Rouleau 2009; Montmain and Penalva 2003).

As modelers, when we proposed to add a social dynamic inside a model of a RSES – such as the inheritance system – we often faced the opposition "we do not have enough data about this phenomenon and should therefore not describe it inside the model (or at least, do not put my name)." The "not-enough-data" assertion is theoretically always valid, and the need for more data should be infinite: whatever the issue, data will be lacking unless one tends to build a 1:1 model! This anecdote illustrates the gap between sciences with different epistemologies, such as those related to the growing interactions between modeling and archaeology.

It is true a model grounded into no data would not contribute the progress of knowledge. However, computational models simulate dynamics and might therefore be irrelevant if a component having a strong influence on the system is not described – in this case, it might be better to introduce hypotheses and question them by simulations rather than build a wrong model because of the availability of data. Behind this anecdote stand a misunderstanding about the role of the model. A computation model is not supposed to represent a complete and definitive theory of the RSES but only stands as a tool to question hypotheses by analyzing their consistency and consequences once extrapolated with simulation.

Beyond the question of data, models should be seen more as a dynamic attempt to formalize archaeological and paleoenvironmental knowledge and also to confront and integrate these systems with perspectives from other disciplines such as agronomy, zootechnics, and socio-anthropology and even more conceptual views on socio-ecological systems (for instance, Janssen and Ostrom 2006a, b). The difficulties of collecting paleoenvironmental and archaeological data in such a way as to compare them and construct a conceptual model of nature/society interactions are highlighted by numerous studies. Some research identified the relationship between the difficulty of explaining the causalities of dynamics and the dependency among assumptions, scale, and forcings (Carozza et al. 2015; Lespez et al. 2016). Modeling appears to be one of the solutions for exploring these complex causalities. Such modeling is thus a means of conceptualizing the dynamics within complex systems as well as serving as a testbed for addressing hypotheses that have been impossible to discriminate and determine which is dominant.

Models of RSES are not definitive proposals of theories but more tools to help researchers to think, communicate, and collaborate. This essential point being clarified, we first start this editorial article by first recalling why these models are built. The art of modeling past societies remains a recent and difficult process paved with hidden constraints, stakes, and issues in terms of epistemology and methodology and is therefore discussed. Finally, we propose a methodological and reflexive process based on elements gathered from several modeling experiences, a mix of good practices shared among practitioners of this field and from our personal experience.

What Models Are We Talking About?

As retraced by Gilbert and Troitzsch (2005), the first attempts to use simulation to study social phenomena started as early as 1960, before intensive explorations in many domains during the 1990s. Models of rural socio-ecological systems were explored during this last, and recent, stream of studies.

Models of rural Neolithic socio-ecological systems attempt to reproduce in computers the sociological entities (households or individuals), the ecological aspects (climate, land and streams, vegetation both cropped and uncropped, fauna, both raised and not), and the relationships among them that occur under certain conditions of time and space. Such a model is necessarily made up of a patchwork of the knowledge of several disciplines. Regarding only Neolithic issues, many modeling attempts have been successfully assessed (Ebersbach 1999; Kohler and Gumerman 2000; Dolukhanov and Shukurov 2003; Janssen et al. 2003; Ebersbach and Schade 2004; Hazelwood and Steele 2004; Janssen and Scheffer 2004; Allen et al. 2006; Kohler and van der Leeuw 2007; Altaweel 2008; Janssen 2009; Lemmen et al. 2009; Tipping et al. 2009; Patterson et al. 2010; Graves 2011; Kaplan et al. 2012; Kohler et al. 2012; Lemmen and Khan 2012; Yu et al. 2012; Carrer 2013; Baum 2014; Saqalli et al. 2014; Lemmen and Wirtz 2014; Bernabeu Aubán et al. 2015), among others, which the study of Saqalli and Baum (2016) sought to characterize according to scale and conditionalities.

This type of model is often built following a spatialized modeling approach with many pixels as pieces of land and many entities, called agents, as households or occasionally individuals: with only pixels, such distributive models are referred to as cellular automata; with agents, they are referred to as agent-based models. In the latter, the various entities and components of interest in the sociotechnical system are explicitly represented in the model. They might be represented with different levels of granularity: for instance, sheep might be represented as individual sheep or as herds; individuals might be grouped into households or individualized. All of these entities are spatialized in the environment, meaning they have a location in the simulated environment, can often move, and are able to perceive and act on it. The environment is most generally discretized and represented on a grid. The behaviors of each entity and of the environment are modeled together and thus at the same scale, for instance, 1 km^2 or 1 ha or more, as well as the interactions among the entities and among the entities and the environment. Entities "live their lives" during the simulation, along the same time pattern and therefore along the same succession of timesteps, of 1 week, 1 month, or 1 year; during the simulation, they move in the spatial environment as they are thought to do. All these models can be considered to be KIDS (Keep It Descriptive, Stupid!) models (Edmonds and Moss 2005), meaning they are literally designed to describe the overall dynamics. Results of such models eventually contradict hypotheses, obtaining surprising or counterintuitive findings as a result of such intricate and complex sets of processes. We position our conceptualization of modeling following van Gigch (1993), Batty and Torrens (2001), Bouleau (2001),

Beven (2002), Couclelis (2002), Kieken et al. (2003), Boero and Squazzoni (2005), Franck and Troitzsch (2005), Lake (2015), Rodgers (2016), and Schulze et al. (2017).

Why Models Are Used for Apprehending Rural Socio-ecological Systems (RSES), Both Past and Present Time Ones?

While the usage of *computer* models might seem to be a novel methodology at first glance, archaeologists already model without computers, meaning they construct hypotheses and theories from collected obtained data. This demarche, or modality of inference, often called "inductive" in the most accepted hypothetico-deductive research, i.e., deduction, induction, and abduction (Blecic and Cecchini 2008), implies the overall combination of "elements" or pieces of science along a plan that may enlighten an issue to be confronted with other facts for validation. In the past, this may have been assessed in an intuitive process, such as that described by Bergson (1911): he described this capacity of connecting processes, patterns, and dynamics along sometimes illogical analogies, comparisons, and consistencies as a purely human action, which is impossible to formalize. As a consequence, a first reason for using computational models is just that as any scientific domain, archaeology produces, encodes, and communicates theories; computational models are just one type of such a model, among others.

A model is a simplification of a system built to help an observer to answer a question on the system (Minsky 1965). Models are encoded using a symbolism (Minsky 1965; Ostrom 1988), which comes with methodological and epistemological benefits and constraints. Models encoded as mathematical equations are not ambiguous, are compact, and can therefore be used to communicate theories easily. Using deduction, they can sometimes extrapolate theories to produce novel knowledge, such as to prove the states a system can or cannot be reached or that a representation of a system is consistent or not. Unfortunately, mathematical modeling has a limited expressive power, especially when it comes to model complex systems made of entities in interaction (for instance, mathematics cannot capture dynamics simulated by even simplistic cellular automata; Wolfram 2002). Discursive models encode knowledge and theories in the form of natural language and have therefore unlimited expressive power but are more verbose and remain as ambiguous to transmit knowledge as human language in general (Eco 1990). On expressive power, *computational models* offer a trade-off between discursive and mathematical models (Ostrom 1988; Taber and Timpone 1996); they can be used to describe systems made of spatialized heterogeneous entities in interaction.

Discursive models do not help scientists to directly generate novel knowledge; once written, the words do not produce novel words which deduct consequences from the written premises; in other words, discursive models allow encoding and communicating theories, but do not produce novel knowledge as a mathematical proof would. If they do not provide any mathematical proof, computational models can be explored by *simulation*, which computes the evolution of the system in time according to the rules which were encoded inside them; this *generative power*, as named by Epstein (1999), was said by Axelrod (2006) to constitute a third way to do social sciences. Simulation produces knowledge because it helps researchers to *discover the consequences of the theories they encoded*. These consequences often might be obtained only by simulation and would not have been reached just by reasoning. Various reasons explain this fact: computers can simulate the consequences of simple rules on many entities, which is beyond the capabilities of the human brain (as for the rules for forecasting meteorological previsions are relatively simple but only can be computed by powerful computers). Sometimes the chaining of the consequences of loops might create complex and nonlinear interactions such as bandwagons, cascades of effects, reinforcement loops, and/or cyclic dynamics such as the well-known Lotka-Volterra dynamics. Sometimes the local behaviors described by the modeler create phenomena of an upper order of analysis (segregation our simple local preferences; traffic jams out of the behavior of cars) which is said to be emergent (Deguet et al. 2006). The second motivation for the usage of models for the study of RSES is thus the fact computation models *generate knowledge* (Epstein 2008), as they *help us to assess the consequences of our theories.*

The interest in modeling tools for collaboration, especially when implemented to integrate both qualitative and quantitative information and rules, is that they are more neutral and balanced, limiting ideologies, preferences, and bias due to the obligation of explicit and written formalization of rules: findings are transparent and can be checked and discussed among a community of researchers (Etienne 2010; Etienne et al. 2011).

Another motivation for the use of models is that it is simply convenient: as stated by Kohler and Van der Leeuw (2007) in their introduction to "socio-natural models," the modeling approach helps to go beyond the narrative description of a society of many archaeologists by avoiding ambiguity and including complex rules beyond simple and deterministic linear connections among elements, usually humans and resources. With regard to this point, we take a social constructivist view of considering societies (for instance, the values, ways, and practices by which a society use is contextually defined, which may vary among societies and within them because they are produced by the society itself, opposed to essentialist perspectives), but as a more essentialist way of regarding physicality, meaning that we consider the constraints that limit and determine rural societies, both directly (such as the ecological productivity of a territory regarding a set of techniques and social practices) or indirectly through, for instance, demography. Therefore, if social laws can be considered in terms of our own mental constructions (unless a right has been written or edited), agroecological laws are "real," meaning that imitating them or following them (according to the epistemological position one may have, either constructivist or essentialist, respectively) is not the purpose of a model focusing on RSES: we take law of biology and physics for granted.

Finally, unlike already-fitted theories, models can improve themselves: as other experiments, they can serve as a step-by-step trial/error demarche. The production of knowledge is not one shot, such as collecting data/creating theory and explaining observations/collecting more data. Experiments with the model also raise questions, contradict theories, or raise difficulties. Actually, we do agree with this: such rules are difficult to settle as each of them may imply many factors and not merely one or two for each side. For instance, the land use of a 1-ha pixel by nearby villagers may be settled in models according to distance to houses, local pedology, local cover, as well as the food requirements and manpower availability of each household using this land according to each household local rights-to-use, each factor having its own dynamic with stochastic variations. Selecting and testing which approach best fits with the external data is a fully acceptable way of using progressive model improvement.

Difficulties in the Construction of a Model of RSES and Outlines

However, because such rules imitate or follow reality, they are all complicated and are thus difficult to construct as a group of variables linked together, then as a parameterized formula, and finally as a parameter-numbered rule. Moreover, one should first define to which entity such formula should be applied: for instance, applying an inheritance rule to a family has different consequences than to an individual – the former introduces the distribution between direct and indirect descendants, while the latter may describe inheritance differentiations between direct ascendants, such as gender or geniture discrimination, which are necessary for all social stratification reconstitution. The same is valuable for the scale of the model, for instance, between a pixel of 1 m^2, 1 hectare, or 1 km^2, which determines the level of details for agricultural activities. The explored issue is then that which determines which scale and which entity level should be selected.

More globally, we see here how complex it is to delineate the issue of what we want to model and for what, before even engaging in actual modeling. The following questions must be applied when building a modeling methodology:

- Can the model answer the questions that are posed? Before that, how can such questions be raised among a community of researchers with various issues, focuses, and questions?
- Can the interdisciplinary pattern and the hierarchy of disciplines be defined? Further, can variables be defined according to criteria that extend beyond the traditional but scientifically questionable in that they are derived from the socially based "gentlemen's agreement" as constructed above?
- Can the model comprehensively explore the various elements to be taken into account while modeling Neolithic rural socio-ecological systems?

Arguments for Rural Socio-ecological Systems' Distributive Modeling

Interdisciplinary Approach: Why Use It?

Several points plea for interdisciplinarity (Porter et al. 2006; Saqalli et al. 2018a) in the RSES study.

"The best complexity": RSES, both past and present, structurally include many disciplines, for instance, one cannot understand livestock-keeping without zootechnics, geography, and anthropology: it is not a question of the methodology to be chosen; it is the object itself that drives the inclusion of these disciplines. Sometimes, such disciplines are eliminated because of a lack of data, which creates objects that are impossible to understand. More precisely, we hypothesize that modelizing all the components of a RSES and simplifying each of them, even unknown, brings less errors, is less harmful in terms of understanding and is by then more reliable than neglecting some elements and focusing on those with many data and/or seeming the most important.[1] For instance, modeling a Lotka-Volterra predator/prey system (Neuhauser and Pacala 1999) without modeling the predator because of a lack of data regarding its ecology would just lead to a population of prey growing as much as resources enable it, without cycles in the population sizes due to the competition between both species. In such a situation, it would be more relevant to integrate a theoretical predator, even if its properties are unknown and require several parameters. One may suggest waiting until enough data are obtained. However, some situations may not allow such a hope: for instance, simulating ancient societies without integrating socio-anthropological rules of inheritance is also useless; however, one may not hope for new information apart for certain clues due to, for instance, differentiated graves.

Systemic approach: More globally, following Verburg et al. (2004), the observations are bound to the extent and resolution of the measurement generated by each observation to provide only a partial description of the whole land-use multiscale system. Beyond the scale of analysis, for instance, the land-use change, it means that scientists and stakeholders must tear down the walls of the disciplinary approach and cultural context that lead to a subjective misinterpretation about such phenomena. For instance, analyzing socio-ecological processes, one of the most complex interdisciplinary scientific objects, combining social and biophysical sciences, the complexity of it must be tackled through the study of the systemic character of reality.

"For the greater good." Indeed, the main quality of social and environmental formalization, and by then a "loud and clear" formalized interdisciplinarity, is to drive scientists and scholars to work collectively to build a common scientific

---
[1] Because interactions are not instantaneous, there is of course not a perfect adequacy among all interacting elements in a single timestep, and the system consistency postulate may be valid only along the simulation and not for one timestep.

object. For instance, reconstituting a livestock herding society implies working together among climatologists, zootechnicians, farming systems' specialists, socio-anthropologists, and of course paleoenvironmentalists and archaeologists, each one "forced" to sacrifice a portion of the complexity of their own themes on the altar of the combination of disciplines. The importance of encouraging the success of the object as a whole and not solely one's task in isolation is crucial as is the legitimacy of the objective and the people involved. The goal of a mediation tool, such as the current model, is to "push" each member to look after the consistency of the interacting system, taken as a whole, such as the example of a livestock-keeping society, for instance, rather than the consistency between one thematic in itself: looking for inconsistencies or even impossibilities within the system is then a good way to test the common understanding over an RSES – for instance, in Saqalli et al. (2014), the impossibility of feeding the quantity of livestock necessary for producing enough manure to keep permanent fields and not shifting fields as suggested by palynology in Linearbandkeramik farming systems allows the research team to propose a systematic pruning practice as the sole practice that will be sufficiently productive.

Exactitude: accuracy vs. precision. Following the previous point, interdisciplinarity is a test for scientific rigor and, more specifically, accuracy. Accuracy is often confounded with precision (Becker 1996), and while precision has this shiny power of data with several figures after the comma, accuracy can be settled only through a reference to a reality, a reference that is difficult to establish independently from these data. But following the metaphor of a target, how useful to send plenty of arrows within a very small range but far from the center? Therefore for instance, what do such figures mean for cases such as, for instance, demographic analyses without including migration? It often happens that scientists present their datasets without justifying the origins of the variables on which and how the data were collected. Why were those data chosen over others? We should then differentiate between variables and data and by then, characterize the scope of the object of research, i.e., its accuracy, before addressing precision. Variables allow relationships and dynamics to be obtained while data parameterize such relationships and dynamics. As a result, we can then define exactitude as the combination of accuracy and precision.

Epistemological formalism. Again following the previous point, these variables must be identified and classified according to a paradigm or a principle (or as the result of a hierarchization and a legitimzation using the perceptions of local experts). We do suggest as a consequence of the formalization of the criteria the principle on which they are designated as relevant. Along the flow back and forth between induction and deduction, there must also be de jure fair criteria for "validation."

Plausibility: Along with this formalization necessity as the second mandatory part of the research process, formalizing elements is protection against self-focusing scientific approaches such as, for instance, the classical self-checking loop mistake where a set of data, for characterizing and "validating" a phenomenon, is compared to the very same data that created it.

We may consider that "validating" a function from one scientific discipline with data from the same discipline has many more risks to create a similar loop. We then plead for steps of "validation," which implies using sets from other disciplines to lower the risks of such loops. The complexity of RSES is not inherent in a specific discipline or a domain. An interdisciplinary approach must avoid all reproduction of domination and should – ideally – transcend the frontiers of each discipline. Youngblood (2007) explains that "what interdisciplinary studies can therefore learn from the bridging disciplines is the importance of not becoming a domain, as domain creates territory and territory creates niche dominance […]." As human beings, researchers act – in a certain way – like nonhuman animals. Ethological studies discuss social animals in a hierarchical community. We have the alpha, the beta, and the omega, which interact and fight for a social position and/or the recognition of liability within a territory. Perhaps researchers should think about "a discipline" outside of disciplinary boundaries."

Why Use Agent-Based Models for the Spatial Reconstruction of Interactions of RSES?

A model is first of all a simplification of something, usually a chosen portion of reality (Minsky 1965). It is designed to answer a limited number of questions (Mazher 2001). The first interest in modeling a dynamic, a territory where societies and territories interact is first of all to agree among researchers from different disciplines on a conceptual model that is fundamentally interdisciplinary with regard to the subject under consideration, which is not obvious. Any model is therefore also a tool for dialogue and confrontation among disciplines (Ducrot and Botta 2009; Maru et al. 2009; Etienne 2010; Etienne et al. 2011).

The spatialized agent-based models used for Neolithic studies thus far have been composed of agents acting over a grid composed of cells, each entity type described by rules, with more or less complex behaviors according to the specifications of the modeling team. The benefits of these models are numerous for reconstituting the interactions between man and the environment in the past:

- Spatialization: Such models provide an account of the territory and its functioning, including the fact that a local combination of various parameters creates de facto favorable or harmful situations that are not obvious (e.g., fertile soils or rivers with no access to water in summer because of karst rocks).
- Interdisciplinarity: Such models are very flexible in answering a question but "oblige" not to neglect environmental or social dynamics without which the model will not work (e.g., rules of commensalism in the case of famine: who eats first? Also relevant is the functioning of a possible transhumance or the collapse of local pastoral resources).
- Entity-specificity: Because these dynamics are formalized at the scale of the acting entity or undergoing the dynamics, the approach is more intuitive for monodisciplinary scientists involved – it is easier to determine a parameter for a family than for a population, the latter parameter being the result of the first

combined with many other variables. However, it allows us to see interactions on a very local scale (e.g., the combination of drought and soils that have become poor, and a small adult population will create local famine and not elsewhere).
- Adapted to qualitatively based low-data issues: These tools, by means of the rules introduced to simulate the behavior of family agents, for example, make it possible to integrate qualitative rules with a significant quantitative importance (e.g., patri- or matrilocality, ultimo- or primogeniture). More generally, they make it possible to simulate the "noise" of societies (the fact that not all rural populations do necessarily do the same thing), and, by using rules based on the literature, the experience of experts in a field makes it possible to manage the quantitative weakness of data, which is the main difficulty inherent in any reconstruction of the past.
- Nothing on the multiscale, multilevel aspect? Do models enable us to link what we know at the scale of the entity with what can be explored at a broader scale thanks to simulation?

Finding the Equilibrium Between Simplicity and Complexity for Modeling Past Societies

- The longer the simulation is carried out over time, the more the simulated society evolves, and therefore the more the model must be generic and increase in abstraction to mimic these evolutions. The same should be applied for cases of ecological variety. Therefore, for more validity, it is better to somehow restrict the genericity of all models, for example, a terrain as ecologically homogeneous as possible and a short simulation time.
- The defects and qualities of simulation models are faces of the same coin: they open up many possibilities but close few. However, a benefit of the models is their efficiency, such as when they are used as an experiment bench and in experimental approaches, over and above epistemological comments and debates, which first of all implies invalidation since one cannot prove that something is true but only that something is false (Popper 1985; Carley 1996; Brenner and Werker 2007; Schutte 2010). However, the more complex and less deterministic a model is, the less we may be able to invalidate something, which is the only way to go beyond gaining confidence in our hypotheses, which is a not a clear-cut gain. Therefore, the less a model is developed in terms of rationality, the better it is.
- Exploring the history of cultures and societies necessarily implies the simulation of many agents and therefore of a large population, first of all simply to obtain significant results. However, the more complex and numerous the agents are, the slower the model will be, and the more likely it is to crash. For practical reasons, model simplification is required to be able to exploit it.

- However, we are stretched between the target of simplicity and an attempt to explore the consequences of complex rationalities. Moreover, the decomposition of these rationalities creates uncertainty about the understanding of the final result. Reducing the rationality complexity may reduce the magnitude of the results but allow their exploration.
- Obviously, we come up against the unpredictable aspect of certain major social movements, such as political conflicts, or major qualitative leaps, such as technical or social innovations, unless we introduce the drivers of these innovations and changes, which will be difficult to establish.

However, simplicity in itself risks bringing nothing out of tautologies:

- Thus, showing in a model that a hydro-agricultural society disappears when water has also vanished is not particularly remarkable. To show that such society could survive there, if only for a while, would be of greater interest because it is counterintuitive; however, modeling the simulation showing this result implies greater complexity in the model.
- Complexity, particularly in the social sciences, allows emergence phenomena to appear. This "small causes, large effects" aspect is often the main contribution of non-environmentally focused modeling. Therefore, a good model is defined according to the target; however, it obtains counterintuitive results due to emergent dynamics because it opens new perspectives and enlarges RSES possibilities.
- The complexity and in particular the precision of the description of phenomena at the interface between society and nature in space and time, such as the stages of the agricultural cycle and their variability, also makes it clear that practices are highly variable and adaptable to environmental variability, but they are also related to past conditions and dynamics, i.e., time inertia. However, almost always, the available data, if accurate, are not at the precision scale of the farmer and his rationality (Alam et al. 2010). Pushing complexity to this relevant scale is the only way to capture this variability and adaptability. Thus, the same simulated culture can thereby adapt itself to several different environments without the need for "forcing" through the introduction of explicit rules.

A Series of Checkpoints Before Modeling Take-Off

For the purposes of clarity, we use "variable" to designate the factor itself ("parameter" can also be used). We adopt this term for its validity and relevancy in answering issues that are addressed in a model as opposed to "data," which we use to designate the numerical values of such variables.

Building the Research Question: OSQHYT

In this section, we aim to formalize the argumentation according to a series of questions. These questions are designed to clarify the purpose of the model and thereby its task. Indeed, we have observed on several occasions that, surprisingly, when model construction is successful and scientific partners see the first simulation outputs induced during implementation, they obtain an extension of their purpose beyond their initial goals. They may even arrive at a distortion of such goals in relation to their most powerful and/or dominant partners. We propose the following acronym for this formalization: OSQHYT.

- Object: What is the territory and/or the population to be implemented and in what order? This will, for instance, define the scale of the model or the level of spatial and temporal precision at which the model should be built. For instance, a model of the Linearbandkeramik (LBK, also named rubaneous culture) culture should clarify whether the aim of the model is to reconstitute the functioning of the LBK village or the dynamics of LBK expansion, which are two different tasks and therefore require two different scales. For illustration purposes, we keep on this LBK example.
- Subject: The model subject defines the part of the object to be explored, distorted, and subject to testing, while the rest should be considered as ceteris paribus. For instance, one should clarify the distinction between territory and society as the first induces a model procedure based mainly on paleoenvironmental data as inputs in explorations on a society with tests based on archaeological data, while the latter implies reliance first on archaeological data as inputs with tests based on paleoenvironmental data. In that case, we may choose to work on the social component of the RSES, with a focus on its spatial adaptability and temporal variability.
- Question: The main issue of the model should be clarified with a question that should end with a question mark and that can be answered with a yes or a no, of course with conditionalities and restrictions. In the current study, for example, one can ask the following question: do the RSES we conceived, including its adaptability based on a sequential rationality following a system of decreasing preference, adapt and correspond to the LBK ecological, spatial, and temporal distribution?
- Hypothesis: This step is the procedure for obtaining a test on our question. In our case, we make the following hypothesis: does this RSES fit with the large variability of archaeological LBK sites and with the century-long presence of certain sites?
- Test: This easier step, once the hypothesis has been obtained, is to propose a formulation of the model's methodology. To avoid the very convenient but less formal temptation of constructing an RSES, which is inconsistent in terms of zootechnics, agronomy, fishery, and agroforestry in terms of its fit with spatial archaeological data, two methodologies can be considered:

- We can build a farming system with environmental preferences according to RSES rationality as deduced from the same era/ecosystem hierarchy of § 3.1.2 and therefore from the archaeological spatial data. We then compare the resulting simulation outputs with the distribution of sites.
- We can propose to select a representative portion of these spatially positioned data to deduce LBK preferences in terms of topography, soils, hydrography, and spatial organization and to test the construction of the RSES based on archaeological data and inferences from other sources following the era/ecosystem hierarchy (see Sect. 3.1.2.) with the rest of the spatially positioned data.

We therefore obtain a complete methodology by addressing the relevant issue until the experimental testing is conducted using our simulation model.

Circumscribing the Model Drivers: AVID – Accuracy of Variables and then Inventory of Data

The problem raised here relates to the recurrent observation that the way in which an RSES-related question is asked is often determined by the availability of data. Thus environmentally deterministic explanations to any archaeological change are often used because it is indeed the only available data. This tendency looks like a chef working only according to what is in the refrigerator: the problem is not posed in such a way as to answer the scientific question as well as possible (regardless of the value of such result) but rather to answer with what one has. This way of doing things is, after all, pragmatic; however, it raises a serious issue: how can we overrule reasoning if we only give priority to the components upon which we have data?

We hypothesize that a primordial[2] qualitative approach, consisting first of defining which variables are to be considered and then obtaining progressively precise results through the evaluation of each of the components of the system considered, can be more scientifically valid in apprehending a scientific issue for which the variables are numerous, as in the case of environmental health issues at the interface between society and nature. This is why we propose as more scientifically valid the search for variables that are essential and primordial before providing such variables in the data once the consensus on the variables to be studied has been reached.

Once the previous point has been set, how does one choose which variables to study and what to measure?

A priori, it is possible to consider testing each variable considered and possibly overturning its importance by setting up an appropriate experimental protocol. However, by listing the variables, do we risk excluding important ones a priori without being able to justify that we were not mistaken? According to Popper (1985), it is precisely not theoretically possible to prove that a variable must be integrated

---

[2] Primordial in its etymological meaning: the primary one

into a problem and can only be invalidated. Thus, what should we do about questions, especially on issues between society and nature, where the number of variables to consider is immense?

A protocol must make it possible to establish in advance the list of these variables to be collected and whose repetition will form the data to be analyzed. The use of variables based on "common sense" or "experience" often fails to "sort" variables by default without specifying how they are selected. In practice, we assume that, in the case of past RSESs, there are only three ways to justify which variables to study:

1. An approach based on its own positioning and its own experience "based" on a more or less recognized expertise, often justified by a publication. This often happens but induces biases.
2. An approach that uses the existing literature on the issue through reference publications usually based on Method 1. Justification is rarely provided for the factors chosen.
3. An approach based on the consensus of the scientific experts' community on the issue, which is technically equivalent to 2). This approach may be formalized through the presentation of a survey or meta-analysis of scientific articles in the field or through formal methods that co-construct the issues (Etienne 2010; Etienne et al. 2011).

Required Qualities of Variables and Related Data: EGI PER PRECIUM

We tend to build a commonly agreed-upon set of criteria that may classify the value of data and sort them, based on the acronym EGI PER PRECIUM (*"I acted according to value"* in a very poor Latin). The definition is initially a first census from Saqalli et al. (2018b), and, although independently conceived, it is similar to that of Pipino et al. (2002) and Batini et al. (2009):

Expressive, Generic, Inter-comprehensive/Perennial data sources, Efficient sensitivity, Robustness/Discriminative, Entangled, Precise, Rustic, Exact, Covering, Integrative, Useful, Measurable. We categorize these criteria in three blocks:

Social and communication usefulness

- Expressive: Variations of this variable should be easily talkative in terms of trend visualization. For instance, the 2 °C level as a threshold for climate change is more talkative than an MW-based representation.
- Generic: This indicator is not field-dependent – it can be constructed from various sources and measurement tools and thereby can be produced from various environments, ecosystems, and study sites.

- Inter-comprehensive: indicators and variables are to be understandable or at least as non-polysemic as possible given the various disciplines involved in the modeling process to avoid misunderstandings.

Sensitivity/robustness

- Perennial in its data sources over time.
- Efficient and discriminative sensitivity: A variable may seem essential, and the corresponding data are excellent; however, if it does not influence the socio-ecological system either in space and time or in the variability of these two elements, or if it influences it but equally and homogeneously, then an equal influence is equivalent to no influence at all, and it is useless.
- Robustness of measurements: The data value is robust and trustworthy regarding the quality of the measurement and/or the operator.

Data and variable efficiency

- Precise: the atomic entity (i.e., the smallest and inseparable unit of the model) should be as small as the constraints of the model allow and as the model issue requires. The atomic entity concerns the spatial grid pixel scale, the temporal rate of time, and the socio-economic survey unit (family or individual).
- Rustic: the variable does not need complex requirements and calculations before modeling and can be used as directly as possible. For instance, precise data, such as pedological horizon heights, should be adapted in terms of the flux to be used.
- Exact: the variables should integrate the complexity of the studied elements and the reasons why some variables in its composition have been neglected should be relevant, which are also valuable for the exactness of the related variability and differentiation according to local differences, such as agroecological conditions. For instance, the food gathering capacity per pixel implies components such as mushrooms, nuts, and fruits. Having good values for only fruits and nuts without mushroom data is less valuable than a rough but closer estimation.
- Covering and complete: the variable should obviously cover the whole modelled territory and the whole simulated period, and no parts of time and space should have zero value.
- Integrative: the variable should allow a simplification by covering a large domain and its value: for instance, building one value for all gathered non-timber forest products is simple as far as it adapts well to seasonal and spatial variabilities.
- Useful, practical: the variable should be quantifiable and measurable. For instance, qualitative rules should transform family dynamic functioning into calculable and modelizable functions
- Measurable: the access to the data should be simple and as free as possible.

Where to Gain Access to an Extinct Society Without Written Documents?

The Relevant Modeling Unit

Several possibilities can be envisaged to simulate human entities, all of which are related to the investigated issue, and, therefore, the scale and the considered functioning regarding migrations and land use are as follows:

1. Individuals: This component allows for the simulation of intra-family tensions and changes and therefore all inheritance transmission, gender or age discrimination, and family organization variations as described above, hence the possibility of considering "cultures" whose adaptation is more or less rapid. However, this requires the formalization of intra-family rationalities on which little information is available; nevertheless, it imposes an enormous number of human entities, each corresponding to a single individual, which is difficult to manage beyond single village levels.
2. Families: This intermediary entity does not allow for the explanation of differential adaptations to the environment by the family organization. However, it can envisage inter-family differentiations on which assumptions about differentiated migration among families can be made. One can consider that this allows an "economy" of entities with a ratio of 1:5 to 9 in relation to the "individual" entity.
3. Villages: We can consider this fixed entity as creating other villages. It is possible to create the attributes "number of families" or "number of individuals" but not "records" of family dynamics, which means that it is impossible to discern families and even less so individuals and thus no differentiation between these entities in their use of the land's resources: it would be the village as a whole that would evolve. However, this scale is relevant on the global scale for entities such as continents. One can consider that this allows an "economy" of entities of a ratio of 1:5 to 20 in relation to the "family" entity.

Rationality and Structure of the Social Component of Socio-ecological Models

Any variable needed for modeling the social component of RSES, past, present, and prospective, can be used to address rationality regardless of whether it pertains (that we may define here as the capacity, always limited, to make choices between practices, activities, and social mechanisms, including norms and rules) or not according to the explored issue. In a socio-ecological system, some variables are based on functioning principles that are homogeneous regarding time: for instance, demography follows equivalent dynamics whatever the era and has the same impact on population growth and structure. On the other hand, some variables do create more uncertainty as they depend more on history of values: rationality, for instance, determines the strategy used by villagers for their farming practice (maximization under constraints, securization, maximal diversification, etc.) and is harsh to deduce: there is nobody to interview to gain access to the rationality of people, and these rationalities should be guessed.

Obviously, as with Hamlet and the skull of Yorick, one may complain about the lack of communication from such remnants: no interviews can be assessed from anthropological investigations; however, the information provided by such remnants is necessary as only such rules can directly characterize family and society dynamics, and these last components are necessary for reconstituting the functioning of societies: what are the marital practices (polygamy or monogamy?) or the inheritance transmission practices? What are the various and differential rights-of-use? What are the rules regarding manpower and resources' organization and affectation within families and between families? What are the colonization practices, meaning, what pushes people to leave? What criteria do they use to choose a new place? Through what ways did they leave their place of origin, through individual families and/or groups? Admittedly, few quantitative data can be obtained in present-time models as well through, for instance, the analysis of socio-economic questionnaires.

Of course, quantitative data are also needed to reconstitute the structure of these societies, such as Gini indices regarding wealth as well as family size and the allocation of resources among subgroups. One may obtain quantitative proxies of some parameters through archaeological indices, such as the number and organization of poles delimiting house size, which is considered to be a proxy for the of people living in the household.

The degree of complexity to be considered for the behavior of human entities depends on the purpose of the model and therefore on the macro-observation scale. Too many interagents allow behavior to be mimicked in a manner that is similar to that assumed but do not allow either explanations of the hypotheses chosen in a manner that is sufficiently short for publication or a full sensitivity analysis given the large number of parameters (Chattoe 2000; Amblard et al. 2006).

Several syntheses, such as those from Axelrod (2006), An (2012), Jonker and Treur (2013), Livet et al. (2014), Malawska and Topping (2016), or Abar et al. (2017), formalize the dilemma between social relations and the complexity of agents' rationalities. The more complex an agent is, the more its behavior must be justified because any rule introduced to optimize a function can establish different forms of "forced" behavior, each applicable to the different functions of an agent. Establishing hierarchical mechanisms can help in addressing the complexity of the problems to be controlled and even reduced. As a *Primum non nocere* principle, we propose to follow a sequential behavior – "People do what they have to do when they have to do it." An example may illustrate these positionings:

1. A model for describing farming systems may be voluntarily restricted to reproducing crop cycles and other resource allocations throughout the year. The objective is to obtain the apparent reproduction of the different actions, by making agents simply reactive: "Farmers and other users use resources at a particular

time and by a particular means," without an assigned objective. They do the things we want, and we observe results at a more global scale to see if it fits with archaeological data.
2. Finding the reason why people were doing such activities in such ways means introducing hypotheses about objectives that are supposed to be used by these actors, their rationality. The point here is to try to imagine why cause-and-effect sequences are used. These objectives can still be made as basic as possible: "Farmers and other users manage their means of production to best achieve their objectives, namely, to eat throughout the year."
3. Exploring the space of parameters of these rationalities, the spectrum in which these rationalities and practices can adapt themselves requires the implementation of prospective scenarios, and in archaeological cases, scenarios mean other sites and environments and contexts (introduction of norms, innovations, social or institutional changes, new territories, etc.) leading to the definition of conceptions further upstream: "Farmers and other users seek to achieve their respective objectives by negotiating with the already settled rules, by practicing various activities and managing them." And so on, the reasoning is refined to move from reaction to cognition, from simple to complex, but also from information directly resulting from data to the translation of hypotheses about behavior and rationality.

Advantages of the first approach are twofold: it is simpler to settle and then to explore, and results are robust facing threshold effects introduced by any artificially assumed optimization function as well as the fact that it is unassailable in terms of data suitability.

Hierarchy Criteria for Seeking Rules of the Social Component of Socio-ecological Models

In any case, for both structure and dynamic variables, the social component of past RSES can only be guessed by borrowing information and by adopting rules from external sources. One should then construct a hierarchy of the validity and legitimacy of these sources; for the current study, we propose a hierarchy based on the proximity to the concerned RSES. Again, we take the view of simulation as nothing but an eternal source of fruitful errors, producing asymptotically improving representations of reality but never reaching it.

For the rules regulating the social component itself, the following should be noted:

1. *Ad antiquitatem* or anteriority: if a family and a collective system had occurred somewhere, it can be considered to be a potentiality for the studied society; however, no information can be deduced from the absence of past anthropological rules. Thus, it is only a principle of probability allowing a past anthropological rule to be more likely present in the concerned society.

2. *Ad populum* or majority of the proximity: if a family and a collective system occurred in a neighboring culture and, even more, in the majority of neighboring cultures, it can be considered to be a potentiality for the studied society; however, no information can be deduced from the absence of this anthropological rule. Thus, and again, it is only a principle of probability allowing a past anthropological rule to be more likely present in the concerned society: it has happened that anthropological rules have appeared and/or crystallized in societies by cultural, ethnical, or class opposition or other restrictive mechanisms artificially imposed by power holders.
3. *Primum non nocere:* social information about past societies is so lacunar and flawed that we propose first to replace "obvious" or classical rules for a more "innocuous" functioning – for instance, choosing patrilineal systems for societies for which no social stratification has been observed through, for instance, funerary, differences, may appear the most evident rule by default. However, it creates such bias by producing social stratification between female and male heirs, thereby creating "naturally" social differentiation that we suggest adopting in the case of ignorance regarding such issues in a bilinear system (Table 1).

For rules regulating the connections between social and environmental components, i.e., the practices and techniques allowing humans to use and transform natural resources, we also hierarchize the reliability according to principles of anteriority and proximity; the latter applied to neighborhood for technologies and to ecological similarity for ecologically constrained factors. We then propose this succession of conditions from the most reliable until the least one:

1. The "best" source is of course the "same era, same territory" situation: Data can then be obtained from directly concerned archaeological sources and extended to the whole concerned territory by inference and generalization. For instance, one may first suppose that Linearbandkeramik (LBK) families were enlarged

Table 1 Formalizing the combination of territory and era factors for determining the validity of data inference

	Anterior era	Same era	Posterior era
Same territory	(3) Manageable environmental and technical packages	(1) Ideal situation	(5) Manageable environmental package. Technical package to be clarified
Nearby territory	(2) Extension based on the hypothesis of local homogeneity		
Other but ecologically similar territory	(6) Tendency for environmental determinism	(4) Tendency for environmental determinism	(7) Tendency for environmental determinism with less reliability than (6) and (7)
Other and ecologically different territory	(9) Quasi no reliability: the technical package may be theoretically acceptable	(8) Quasi no reliability: the technical package may be theoretically acceptable	(10) Nearly useless

multinuclear because of the multiroom elongated shape of their houses compared to contemporary cultures and then extending this family system to the whole culture.
2. "Same era, nearby territory": extension can also be used for technical capital based on the postulate that rural farming or hunting/gathering societies living nearby for enough time have access to an equivalent set of technologies – we can then suppose that the availability of a technology in nearby cultures may allow the presence of this technology within the concerned culture; however, the absence of a technology in all neighboring cultures suggests that it is less likely to see this technology in the concerned culture.
3. "Anterior era, same territory": many exceptions do occur in the temporal progressivity of technical capital along history; however, the trend is largely in favor of temporal extension allowing posterior inclusion of previous practices and techniques, at least within the panel of possibilities available for simulated humans.
4. "Same era, other but ecologically similar territory": We use the term "ecologically" by integrating the manpower ratio compared to ecologically constrained needs, implying, for instance, that highly manpower-demanding weeding and watering steps in the farming cycle may be equivalent for both Khmer and Mayan forests, which can inform both the panel of practices and techniques available for the concerned society with a lower reliability than 2 and ecological "determinism" with a higher reliability than 2 if the latter concerns different but neighboring ecosystems.
5. "Posterior era, same territory": to avoid anachronism, extension may be defined especially regarding ecological constraints according to the difference of technical capitals between the two periods, including social innovations such as manpower restrictions or, conversely, collective manpower mobilization – for an equivalent capacity of transformation of the territory, one may then use information such as a fruitful (and sometimes apparently obvious) restriction, for example, arid areas in the Middle east, even those close to large Mesopotamian rivers, were not irrigated beyond a certain extent even with the appearance of energy-multiplying techniques. Thus, socio-ecological models should consider areas such as those that are absolutely non-irrigable. However, we consider it to be less reliable than 4 because the diffusion/innovation of techniques has significantly more impact on the relations between society and nature than the variability of these relations among societies.
6. "Anterior era, other but ecologically similar territory": no information can be here deduced regarding the panel of practices and techniques available for the concerned society; however, such information may be helpful for the ecological semi-"determinism" regarding the use of natural resources. For instance, some practices regarding livestock in open territories, such as steppes and savannas, can be used, such as herding (how many animals for a shepherd), but not all of them, such as prolificacy.

7. "Posterior era, other but ecologically similar territory": the argumentation here is equivalent to 6; however, for the same reason as in the difference between 3 and 5, it is far less reliable.
8. "Same era, other and ecologically different territory": here, the sole element to be integrated when no information is available is the maximum panel of technologies and practices to the extent that the two considered cultures had contact. If not, nothing can be said.
9. "Anterior era, other and ecologically different territory": again, it is equivalent to the previous level, with less reliability.
10. "Posterior era, other and ecologically different territory": nothing can really be said.

Let us not forget that if we assume isotropy of rationality and ecology, i.e., that the present behavior of the system is related to the past, then we need to update the meaning of the available information, and it requires a sustained analysis and synthesis work and much reflection as the conditions under which the facts occurred were not, necessarily, in the same context in which they now occur.[3] Then, if we want to use the information of the present and the past to propose a future, then we must resort to forecasts whose methods have limitations that require such information to be conditioned. Behind all of this is the human being who directs the thought to build methods and models of information processing.

Modeling Take-Off: Piloting Tricks

Initialization: Avoiding Initial Distortion

Agent-based models are temporally defined. At $t = 0$, whatever the length meaning of the timestep, and as many functions are evolution processes of the same simulated variables, these lasts should be initiated. Posing it a priori creates distortions and wide variations in the outputs of the first timesteps, especially if their values are very different from the variable average values. One may propose, as happens in KISS approaches (Edmonds and Moss 2005), to wait for some timesteps by considering them to be nonvalid; however, this distortion may remain even if the related distortion cannot be detected through, for instance, too large impacts on effect-accumulating biophysical factors, such as fertility. Finally, and more conceptually, history is a permanent process with no beginning; thus, seeing unhistorical fluctuations is very depreciative for the outputs' appearance from the point of view of thematicians. We then propose to reduce as much as possible the corresponding

---

[3] For instance, even if ecological conditions at the beginning of the Holocene were suitable in central Europe for many tree species present now, some species were absent due to the fact that they are actually alien, having come from the Americas or Asia, or they may have only been slowly recolonizing the continent from Mediterranean shores, which takes time.

fluctuations by assuming, for instance, to calibrate initialization values as simple means deduced from the first simulation sets. One may then suggest measuring the stability of the solutions and designing a mechanism that helps control it before using the model for further explorations.

"From the Top of This Pyramid, Forty Centuries Look upon You" (Bonaparte 1798)

Courdier et al. (1998) describe the construction of a model as a spiral where modelers come back and forth on the various modules of the model through progressive adjustments. We adhere to this point of view; however, we consider that this can be reduced through the hierarchization of modules because of inter-variable dependencies, from which the incidence of such dependencies should be evaluated. These dependencies are nested or more precisely structured in a pyramidal way (Fig. 1). At the base of this pyramid of dependencies and therefore as the first basement bricks of the pyramid to construct, one should build independent abiotic factors, such as climate, topography, soils, river and shoreline movements, as well as the range of present plant and animal species, from which we deduce their distribution and their spatial and temporal variability, the diachronic fertility of soils. On the next floor, and both deduced from the latter and combined with fixed rules, social structures, and technical capital, we then construct the practices of natural resource uses (agriculture, animal husbandry, hunting, gathering, and fishing). On the next floor again, the dynamics affecting societies, affected by production activities' differential efficiencies, are elaborated, as much on the round run, the annual and generational cycles. Finally, on the last floor as the long run, one may complete the pyramid with the dynamics of stratification and separation/colonization.

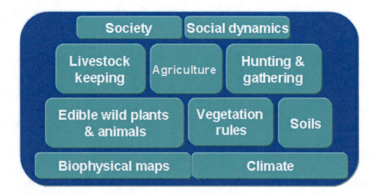

Fig. 1 The pyramid of dependencies

The Modeling Process: An Experimental Demarche

The model is only useful if it reflects not only the situation that one wishes to mimic or simulate but also the space of alternatives, reduced to a minimum to simplify the task but sufficiently broadened to answer the question at the origin of the model: only the robustness of a simulated answer along various parameter alternatives can show that this answer can be validated, even with fragile data.

It is then necessary to have confidence in this model, and this is only possible through a confidence-building test (not a real validation as it cannot exist, as described above) first through a calibration at the micro-entity level ("does it work the way we want?"), then at the macrolevel ("do the environmental settings behave like we need?" "does the population of this village behave along the constraints we planned?"), and then, the most important, the comparison with external data not used in the model (Fig. 2 *left*). Only then can the model serve as the bench of experiments, the test-bed to crush, modify, and tinker with the use of scenarios (Fig. 2 *right*). We therefore make a call for modeling as an experimental approach.

The scenarios described as the final but most proficient phase of modeling presented in Fig. 2 right must fulfil three steps:

- As part of the confidence-building steps, each scenario is actually an exploratory distortion of the average "business-as-usual" base scenario. However, as such models are always complex and are actually growing in complexity and heaviness over time, including many functions and many variables (Rubio-Campillo 2015), a full sensitivity analysis soon becomes enormous following an exponential law. For instance, assessing a complete single-parameter sensitivity analysis on a model with 52 variables implies, with a minimum of 5 degrees of freedom and 20 simulations each, 5200 simulations. We therefore use a partial sensitivity analysis on the most relevant factors, according to the potentiality of the variability by which they occur, their weight in the model, and the panel of the scientific disciplines they cover. One may then suggest using the faster Morris methodology (Morris 1991) which provides parameters' tendencies and qualitative rankings but cannot be used as a screening method (Ye and Hill 2017).

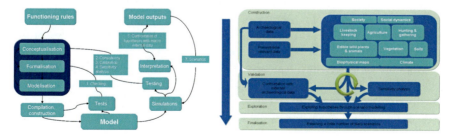

Fig. 2 Elaborating a modeling methodology

- The base scenario itself is to be explored once confidence has been obtained through outputs comparison with non-included data by exploring simulation outputs on areas where no data have been acquired, for instance, in archaeologically unexplored zones: discovering remnants following simulation outputs is the best possible confirmation of such a model. Other ways include instantiating the model in a different geographic configuration and a different but similar culture to extend its genericness.
- The scenario exploration part, which can be considered to be a retroprospective demarche (Paegelow and Camacho Olmeido 2008; Verhagen and Whitley 2012), may follow two demarches:
    - The *reductio ad absurdum* exploratory purpose: this implies using the model with voluntarily caricatured scenarios to invalidate hypotheses. Usually, such hypotheses are non-nuanced ones, for instance, climate-deterministic scenarios. Simulations may show that one simple reason is insufficient to explain the dynamics implying the acknowledgment and consideration of multifactorial combinations as often occur.
    - Finally, one may then use the model as a comparison among various legitimate hypotheses to determine which of the scenario simulation outputs best match the field data: no validation can be provided in that way; however, a plausibility-based hierarchization can then be obtained.

Integrating Data: Part Margin of Error, Part Variability

Every piece of data obtained through measurements is entailed to a margin of error, especially if it concerns past issues. Even more, paleoenvironmental or archaeological data are rare, and a statistical analysis on the variability of their values is thus difficult, apart from some repeatable data, such as the number of house poles or non-ritual everyday pottery patterns. For instance, palynology data providing temperature and rainfall reconstitutions from European Pollen Database sources (http://www.europeanpollendatabase.net/index.php) have a margin of error on these two variables. Transforming such data into adapted ones, i.e., at the month level and for each km^2 of a territory, implies creating random monthly temperature and rainfall, necessitating a standard deviation for doing so. We thus used the margin of error as a maximum variability for a proxy of the standard deviations we needed, greatly increasing the range of extremes of the reconstituted climate. Such a rough procedure may be considered only in the invalidation-based experimental demarche through the caricature-then-nuance procedure described above. This point can be considered to be useful for elements of the social components of RSES, such as family size (using house size as a proxy), livestock size (using meat consumption as a proxy), or gender inequality (using grave wealth as a proxy).

Social Modeling Difficulties

The Scale Gap

Several examples of archaeological modeling face harsh issues when they must combine at the right scale the interactions among social elements. Indeed, the social component of RSES altogether implies the following:

1. The evolution of the technical and practical assets, i.e., all the components parameterizing all production and economic activities practiced and used by the concerned population for living; it then defines the differential productivity according to the systematically most rare resource, the manpower (except very rare preindustrial situations, such as ancient Egypt, where land was even rarer).
2. Data characterizing the demographic growth and variability according to food availability, the latter being defined due to economically, ecologically, and socially related productivity.
3. Rules defining the conditions of accessibility of all family and group members to the different production activities and related products (various food), thereby defining the everyday hierarchy within human groups.
4. Rules of stratification including marriage, inheritance transmission rules, and rights-of-use that allow the creation of inequalities are important points as they can be compared to the social stratification of usually better conserved and less taphonomically altered graves as well as altering the most equal and more optimal use of manpower and gender distribution (for instance, through polygamy).
5. Rules of family and group splitting and geographical movements, defining the power of expansion of a village and its sensitivity to ecologically defined parameters through the temporally and spatially variable manpower productivity of all production and economic activities as well as the influence of non-ecological factors in the choice of new settlements, such as the force of the habitus, i.e., the variable definition of the preferred oekoumen according to the past history of the concerned social group.

Putting together all these elements in a single model is a challenge because they do not forcedly correspond to the same spatial scale: for instance, the level 1 may correspond to the culture as a whole and then, going deeper at the village level, one should integrate discrimination regarding the mastery of this technology (for instance, livestock herding). The level 5, because it should include both the departure and the arrival areas, may be implemented at a larger scale than the others. Anyhow, the more we implement social factors, the more diverse the villager agents should be to cover all the socio-ecological situations in the model, the more complicated the villager agent should be, the more the model requires time for implementation and simulations. One may then observe that agent-based models focus on some of these elements: for instance, the model of Premo (2006) on hunting-gathering Paleolithic groups focuses on the point 3, in this case the commensality within a group of humans. Most models can be classified into categories according to the scale that is itself related to the issue but also the underlying hypothesis (Saqalli and Baum 2016): we then separate models into local-scale models and large-scale models, which we describe through a non-exhaustive series of some examples:

- Within the local scale (i.e., the village or the group of villages), the famous founding work of Kohler, Axtell, Epstein, Janssen, and others (Janssen et al. 2003; Janssen 2009) on the Anasazi give priorities to points 2 and 3. The water constraint is so harsh that the resulting situation is de facto binary: if there were no water, the Anasazi people would not be able to survive. An equivalent model (Baum 2014) explores the various practices regarding land use at the village level, thereby connecting 2 and 3.
- At the global level, GLUE (Wirtz and Lemmen 2003) explores the transition to the Neolithic era through a transmissible set of techniques expanding through colonization and diffusionist patterns through the Ancient World, with a focus on certain areas and exploration of climate deterministic hypotheses (Lemmen and Wirtz 2014), which we can translate into a combination of 2 and 5, explored through scenarios based on 5. The model of Conolly et al. (2008) analyses the diversity of the set of cropping plants and connects it to environmental perturbations and diffusion, meaning combining 1 and 5. The model of Bernabeu et al. (2015) addresses Neolithic expansion patterns, also combining 2 and 5.

A project (ANR Obresoc) with two divergent objectives, reconstituting the farming system of the LBK and the expansion then decline of this culture across time and space, was even forced to split it in portions with one focusing on the livestock and farming systems (Saqalli et al. 2014) and another on 5, the expansion itself.

As a consequence, we then call for the formalization of the positioning of the social component of the issue along at least two of the 5 points presented above, using one as an object of research and the other one as a subject to be explored (see Sect. 2.1).

Social Component Affecting Binaries

This section tends to demonstrate the powerful impact social elements may have on modeling outputs:

- Inheritance and gender stratification: first, one may have to note that inheritance in a low demographic density context (regarding the density in relation to fertile lands) concerns only mobile assets, i.e., livestock, the large availability of lands reducing to zero the value of such lands. Whatever the gender-defined discrimination regarding inheritance, whether female or male heirs are advantaged, it creates de facto as social stratification which extent depends on the force of this discrimination. Let us suppose that this discrimination advantages males and is absolute, i.e., there is no part of the inheritance given to daughters and no dowry as compensation. Therefore, families of only daughters must distribute in other families their assets (through a lineage link or randomly) and cannot conserve a patrimony. However, sons-only families conserve their assets. A second case proposes a non-absolute discrimination (for instance, 2/3 for sons and 1/3 for daughters): with patrilocality married daughters bring their assets out of the family patrimony, "disadvantaging" their native families regarding patrimony conservation and advantaging their new ones. Gender natural variation may be considered as balancing the dynamic to keep equivalent gains and losses among families. However, luckily "advantaged" families and lineages, by having (still in our example) sons, both accumulate assets from spouses and create new family branches providing security to the whole lineage, compensating thereby the non-patrimony effect. Therefore, initial random differentiation keeps some lineages advantaged, and distribution of transmission rights should be carefully under consideration: Todd (2011) proposes as a prior system before the great extension of the patrilocality a system of origin based on bilocality and bilineairity.

- Ultimogeniture vs. primogeniture: an equivalent discrimination phenomenon occurs between elders and juniors during transmission if a differentiation is defined. Moreover, this has an important effect on the power of expansion of a culture: Let us consider a theoretical case of families with all two children. The elder family has kids statistically older than the ones from the junior child. If this culture uses ultimogeniture, meaning it is the elder child who leaves, his/her children will come to age before the ones from the junior family and the manpower capacity of this family will be more rapidly higher. If there is a difference of 5 years between the two heirs, the child of the elder will come on age 5 years earlier. Each generation needing more or less 30 years, this implies a speed difference of 17% with only this anthropological difference.

- Enlarged families/mononuclear families (Radja 2003): this opposition is often forgotten as a major factor of reduction of the manpower constraint and thereby RSES productivity. We show in Saqalli et al. (2010a, b, 2013) that the enlarged family configuration allows families to overcome manpower constraints thanks to family solidarity, for rapidly building houses, clearing new fields, hunting, fishing, gathering, and keeping a large livestock herd at the same time, for example. Mononuclear families are lowered in their expansion because of theses everyday constraints. However, we also show in the same articles that enlarged families are less robust to environmental shocks and variations, and local famines have then far more consequences on demography. From then on, we do hypothesize that a major explanation for the difference between LBK families and equivalent contemporary and neighboring cultures such as the Starčevo culture can be seen in the configuration of houses: the long houses of the LBK are a reflection of enlarged households with many nuclear families, while the Starčevo houses reflect the nuclear family type.

- Practices and geography: this relation leads to various results in terms of one activity efficiency and in pressure on resources according to its geographical configuration. For instance, we have shown in Van Vliet et al. (2010) that hunting through pathways creates de facto a game haven. As other examples, Rouchier et al. (1979, 2001) and Maury (2004) have simulated the daily or yearly transhumance movements of herds at very local or at a medium range, showing topology plays a large role on the distribution of the pressure over land and resources.

Round Time Functioning: Production Activities

For all the economic activities linked to local resources (hunting, fishing, livestock, agriculture), the functions linking initial environment potential (such as fertility and climate productivity), human transformations (such as land clearing, burning, and livestock-based fertility transfer), and cultural operations (soil preparation, amendments, seeding, weeding, harvesting) are to be characterized to produce yields varying spatially and fitting with literature estimates (Rowley-Conwy 1981; Gregg 1988; Mazoyer and Roudart 1997; Ebersbach 1999, 2004; Rösch et al. 2002; Kreuz et al. 2005; Bakels 2009; Malézieux et al. 2009). The various economic activities envisaged are described below.

Wood Needs and Cuttings

Firewood harvesting is the least frequently discussed resource use practice, although according to Mazoyer and Roudart (1997), it requires a considerable amount of land, besides farming and pasture lands. However, as the only practice to be modeled is wood harvesting or cutting, modeling is not very complex. A program specifically dedicated to firewood (Antona et al. 2005; Bacaër et al. 2005) explores scenarios on timber harvesting in Sahelian Africa. Timber cutting is more complex: it must meet the specific demand for houses (or pirogues) and requires specific needs (some species specific to carpentry, a forest old enough and in a suitable place to shelter large trees, etc.). To our knowledge, no modeling has integrated this aspect, but its modeling, if considered justified, does not raise fundamental questions.

Agriculture

In addition to investigations on past agricultural systems, of which there are some examples such as Baum (2014), many models of nonindustrial current farming systems may be used to analyze past rural populations and areas. They are more or less detailed and based on the hypothesis of an optimization of available resources, in particular manpower and land. As mentioned above, these simulations balance between models based on functions derived from correlations of available data and models in which processes are simulated. The former focus on spatial hypotheses about the evolution of these agrarian systems and therefore work on larger territories, while the latter focus more on the impact of social factors, most often on small territories.

The former, when used to simulate phenomena that are highly spatialized and relatively independent of economic and/or social factors, allow a good reconstruction of the phenomena considered, more like GIS integrating cellular automaton functions (Lieurain 1998; Berger and Schreinemachers 2006; Henry et al. 2003). Gibon et al. (2009) have described and simulated environmental responses to farm-level agrosylvopastoral practices at the village territory scale. The LUCC community, from which some examples are Lambin et al. (2000, 2001), Parker et al. (2001), or Chang-Martínez et al. (2015), summarizes the possibilities for understanding the co-evolution of vegetation cover and land use (Munroe and Müller 2007). In particular, it stresses the importance of nonspatial factors (particularly non-biophysical) as determinants of this evolution, such as institutions. Generally speaking, the more the model is oriented toward the analysis of hypotheses involving a large territory, the less detailed the agricultural practices are.

As an intermediate step toward more social and de facto approaches in smaller territories, Bonaudo (2005) developed a multi-site, multi-activity model that introduced the arrival of new migrants along Amazonian routes and the evolution of their multi-crop-forest-breeding production systems during colonization as well as different family factors. The timestep is annual, and agricultural operations are reproduced in very simplified ways. Castella et al. (2005, 2007) and Castella and Verburg (2007) discussed different scenarios for the evolution of Vietnamese mountain production systems based on family agents operating on several types of land and in several workshops. Finally, Saqalli et al. (2010a, b, 2013) simulate several Sahelian agrarian systems based on the combination of polyculture and animal husbandry on a timestep of 1 week, allowing for the development of all the cultivation steps (the shortest cultural step being the sowing, of 1 week for a 1 ha plot, it is this timestep that was chosen) and, in particular, the revealing of link between the sequential management of the agricultural cycle with temporary labor constraints within the farm and temporally and spatially highly variable environmental constraints (rainfall delay, delayed sowing, rapid evolution of fertility).

Livestock-Keeping

All the previous systems included livestock components to varying degrees depending on the nature of the corresponding production system. The chosen complexity or simplicity of the animal husbandry cycle is equivalent to the corresponding agricultural system. We refer to the previous section for work on multi-crop-livestock production systems. Several studies have focused specifically on animal husbandry:

Livestock herds can be represented in the form of flocks or individuals, or they can be represented virtually through their impact on pastures and contributions to the farms to which they belong (Landais 1992; Bonnefoy et al. 2001). The interest in herd representation (family or linked to a "house," e.g., in the case of villages simulated in this way) is simplicity: the herd agent can have specific attributes, for example, "number of individuals for each species." Rouchier et al. (2001) explore the relationship between gifts and counter-gifts (Alam et al. 2005) to define transhumance systems not only governed by environmental rules (time and spatial adequacy between grazing quality and herd needs) but also by marriage and gift-based alliances. This approach is similar to that of Gibon et al. (2009), Le Page et al. (2013), or Bommel et al. (2010, 2014).

Several authors have detailed the herds, either to simulate the shedding of these herds (Maury 2004) or to represent the complementarity of different species on the evolution of the landscape (Mechoud et al. 2000) or to detail the differentiated evolution of herds according to families (Saqalli et al. 2010a).

Livestock herds can be differentiated into species as required. Thus, Saqalli et al. (2010a, b) separated the three species present on the Nigerien Sahel (goats, sheep, and cattle) by characterizing them in terms of harvesting from plant resources (shrubs and herbs) and their specific characteristics (mortality, mortality of young, fertility, resistance to reduced rations, growth but also genus). Similarly, Mechoud et al. (2000) detailed the behavior of cattle and horses, after detailed GPS monitoring in real conditions, to analyze the complementarity of their pastures in mountainous estivas.

The interest of differentiating species and characterizing them in their adaptation to their environment can be linked with the will to reveal changes in the distribution of these species between villages according to their characteristics. This differentiation raises the question of the level of detail desired to characterize the link between herds' dynamics and environmental resources (water source, herbaceous, leaves) as a system. It may include the transformation of both vegetation and herds, the adaptation of pastoral practices as a consequence. It may also involve transhumance, either latitudinally (Saqalli et al. 2010a) or by altitude (Mechoud et al. 2000), which implies simulating practices outside the village border, sometimes very far away.

Finally, the questions asked to simulate the domestication of livestock are equivalent to those for plant species: the appearance of domesticated, locally or "imported" species, the diffusion of this domestication, and the evolution of the species.

Hunting

Mathevet et al. (2003) make the link between the waterfowl population and hunting rights on the Camargue as well as among hunting pressure, availability of rural households, and the hunting population. A series of studies on the Cameroon-Gabon forest was assessed, initiated by Bakam et al. (2001) on the co-evolution between hunting pressure and wildlife hunted by Petri networks, and pursued focusing on several factors in modelling "real" dynamic systems (probability networks, emergence effects and spatial discriminations) as shown in Bousquet et al. (2000) and Van Vliet et al. (2010). The first article simulates the impact of trapping based on traps that are moved regularly, while the second simulates the effect of a type of hunting called "hunting in front of oneself" or "meeting hunting" along family trails.

Other collective (e.g., beaten) or individual (e.g., hunting) hunting practices remain to be considered.

The question of the link between environmental resources and wildlife population dynamics is equivalent to that of livestock. The spatial aspects of wildlife sustainability are of course accentuated for large fauna that require larger spaces.

Fishing

Several works have been developed to represent the functioning of a fishery (Bousquet 1994) or even to manage the fisheries' resources affected by this fishery, such as Soulié and Thébaud (2006). However, it appears that few modeling tests are available to simulate the fisheries-resource relationship itself outside Bousquet (1994). However, it may be possible to simulate the different types of fishing (line, trap, net) according to the same principles as hunting (see the section above).

Social Factors Among Research Community as a Conclusion

We did not insist on the requirements of the agent-based modeling community for formal protocols tending to be the reference methodology in modeling such as the UML formalization for mutual comprehension (Rumbaugh et al. 2004) and the ODD protocol for publication (Grimm et al. 2006, 2010), nor some very common rationality modeling methods such as the Beliefs-Desires-Intentions (BDI) (Georgeff et al. 1999). We consider that these methods should be taken for granted as a convention within the research community.

However, our own observations support the idea that interdisciplinarity is not socially easy for scholars (Turner 2002; Henrickson and McKelvey 2002). Any research seeks a balance between data collection and formalization (without necessarily going through a computer version). The use of computer models is, before any simulation results, an exercise in formalizing these conceptual models. This exercise comes up against two points:

1. They are more easily shared when they are limited to one discipline, first of all because of difficulties in the dialogue between different disciplines: the polysemies and the implicit practices of each discipline are all sources of their own confusion. Many research projects have experienced serious difficulties as a result of this misunderstanding, which is obvious and normal and cannot be solved by simple goodwill: by default, we do not understand each other. Some methods make it possible to formalize relationships such as the ARDI method (Etienne et al. 2011) but involve time and can take up to a full-time week.
2. These difficulties are reinforced by the tacit integration of the limits and approximations specific to the practical exercise of each: we understand each other better between geographers when we speak of spatialized models with only

simplified neighborhood dynamics, and approximations are recognized as normal. We understand each other better between hydrologists on hydro-models whose approximations on underground flows are recognized within the community but that we do not wish to have to explain in other communities. The same goes for archaeologists and paleoenvironmentalists. Hence, there is a need to seek a compromise and not a consensus. The latter is becoming more inaccessible as more partners are included in research projects (the more people there are, the higher the conflict emergence probability will be). According to Nachi (2006), "a compromise is a process that develops between partners seeking to reach agreement at the price of some accommodations, modifications, and reciprocal concessions between competing interests." This relates to the question of domination based on financial, academic and institutional, or affective power among the concerned research community, which should not be avoided but should preferably be formalized. More practically, publishing in one's discipline is socially, academically, and professionally recognized: publishing elsewhere is a low reward effort, especially for disciplines whose associated journals have high impact factors.
3. Finally, more conceptually, no model combining several disciplines is a direct transposition of each of them: a global architecture is necessary for the coherence of the whole. However, this necessarily implies a simplification that is difficult to legitimize for each of the thematicians in their own community when publishing an article, which further reduces the value of involvement; even more, one cannot avoid the involvement of competent thematicians by a working method, such as "show your model, we will tell you if it is good": regardless of how much time a modeler devotes to a particular discipline component to include it in a model, the time wasted and also the probability of error are much greater – the experience and knowledge of a domain by a discipline specialist are fortunately irreplaceable.

The models and the simulation are not cold instruments of prognosis since they reflect the human and professional qualities of those who formulated them and of those who use them. A model reflects the desire for knowledge and curiosity of the specialists and the simulation, and thus there is a need to consider what the model wants to express.

There is a palpable cognitive impasse. Many researchers (reinforced by media image) perceive science to be a collection of data, which comes back to the original idea of cultural cognition: According to the Cultural Cognition Project (http://www.culturalcognition.net/), science "refers to the tendency of individuals to conform their beliefs about disputed matters of fact (e.g., whether global warming is a serious threat; whether the death penalty deters murder; whether gun control makes society more safe or less) to values that define their cultural identities."

The difficulty is in building a true transdisciplinary approach beyond the emotional aspect. The latter has an impact on the question of the research question, the subject, and the object of research: in short, the ability to question everything.

References

Abar, S., Theodoropoulos, G. K., Lemarinier, P., & O'Hare, G. M. P. (2017). Agent based modelling and simulation tools: A review of the state-of-art software. *Computer Science Review, 24*, 13–33. https://doi.org/10.1016/j.cosrev.2017.03.001.

Alam, S. J., Geller, A., Meyer, R., & Woerth, B. (2010). Modelling contextualized reasoning in complex societies with "Endorsements". *Journal of Artificial Societies and Social Simulation, 13*, 6.

Alam, S. J., Hillebrandt, F., & Schillo, M. (2005). Sociological implications of gift exchange in multi-agent systems. *Journal of Artificial Societies and Social Simulation, 8*, 5.

Allen, E., Falconer, S., Hessam, S., Barton, M. C., & Fall, P. (2006). *Modeling of agropastoral human activities using agent-based simulation*. New York: Springer.

Altaweel, M. R. (2008). Investigating agricultural sustainability and strategies in northern Mesopotamia: Results produced using a socio-ecological modeling approach. *Journal of Archaeological Science, 35*, 821–835. https://doi.org/10.1016/j.jas.2007.06.012.

Amblard, F., Rouchier, J., & Bommel, P. (2006). Evaluation et validation de modèles multi-agents. In F. Amblard & D. Phan (Eds.), *Modélisation et simulations multi-agents: Application pour les sciences de l'Homme et de la Société* (pp. 103–140). Paris: Hermes.

An, L. (2012). Modeling human decisions in coupled human and natural systems: Review of agent-based models. *Ecological Modelling, 229*, 25–36. https://doi.org/10.1016/j.ecolmodel.2011.07.010.

Antona, M., Bah, A., Le Page, C., Mahamane, A., & Aboubacar, I. (2005). Using multi-agent modeling for policy research: The case of fuelwood policy in Niger. In CABM-HEMA. *Conference on Multi-agent modelling for environmental management*, Bourg Saint Maurice-Les Arcs, France.

Axelrod, R. (2006). Advancing the art of simulation in the social sciences. In R. Conte, R. Hegselmann, & P. Terna (Eds.), *1999. Simulating social phenomena* (pp. 21–40). Berlin: Springer. Updated version in Rennard J.P. (Ed.), *Handbook of research on nature inspired computing for economy and management*. Idea Group, Hersey, USA.

Bacaër, N., Bah, A., & Mahamane, A. (2005). Fuelwood harvesting in Niger and a generalization of Faustmann's formula. *Comptes Rendus Biologies, 328*(4), 379–385.

Bakam, I., Kordon, F., Le Page, C., & Bousquet, F. (2001). Formalization of a spatialized multi-agent model using coloured petri nets for the study of a hunting management system. In J. L. Rash, W. Truszkowski, M. G. Hinchey, C. A. Rouff, & D. Gordon (Eds.), *Formal approaches to agent-based systems. FAABS 2000. Lecture notes in computer science. 1871*. Berlin, Heidelberg: Springer.

Bakels, C. C. (2009). The first millennia of agricultural landscape. In C. Bakels (Ed.), *The western European loess belt* (pp. 89–98). Dordrecht: Springer Netherlands.

Batini, C., Cappiello, C., Francalanci, C., & Maurino, A. (2009). Methodologies for data quality assessment and improvement. *ACM Computing Surveys, 41*, 16. https://doi.org/10.1145/1541880.1541883.

Batty, M., & Torrens, P. M. (2001). Modeling complexity: The limits to prediction. *CyberGeo, 201*. https://doi.org/10.4000/cybergeo.1035.

Baum, T. G. (2014). Models of wetland settlement and associated land use in South-West Germany during the fourth millennium B.C. *Vegetation History and Archaeobotany, 23*, 67–80.

Becker, H. S. (1996). The epistemology of qualitative research. In R. Jessor, A. Colby, & R. A. Schweder (Eds.), *Ethnography and human development: Context and meaning in social inquiry* (pp. 53–71). Chicago: University of Chicago Press.

Berger, T., & Schreinemachers, P. (2006). Creating agents and landscapes for multi-agent systems from random samples. *Ecology & Society, 11*, 2.

Bergson H. (1911). *L'Intuition philosophique*. 5-11/04/1911, 5th Congress of Philosophy, Bologna, Italy.

Bernabeu, A. J., Barton, M. C., Gordó, P. S., & Bergin, S. M. (2015). Modeling initial Neolithic dispersal. The first agricultural groups in West Mediterranean. *Ecological Modelling, 307*, 22–31. https://doi.org/10.1016/j.ecolmodel.2015.03.015.

Beven, K. (2002). Towards a coherent philosophy for modelling the environment. *Proceedings of the Royal Society A, 458*, 1–20.

Blecic, I., & Cecchini, A. (2008). Design beyond complexity: Possible futures—Prediction or design? (And techniques and tools to make it possible). *Futures, 40*, 537–551. https://doi.org/10.1016/j.futures.2007.11.004.

Boero, R., & Squazzoni, F. (2005). Does empirical embeddedness matter? Methodological issues on agent-based models for analytical social science. *Journal of Artificial Societies and Social Simulation, 8*, 6.

Bommel, P., Bonaudo, T., Barbosa, T., Bastos da Veiga, J., Vieira, P. M., & Tourrand, J.-F. (2010). La relation complexe entre l'élevage et la forêt en Amazonie brésilienne: Une approche par la modélisation multi-agents La relation complexe entre l'élevage et la forêt en Amazonie brésilienne: Une approche par la modélisation multi-agents. *Cahiers Agricultures, 19*, 104–111. https://doi.org/10.1684/agr.2010.0384.

Bommel, P., Dieguez, F., Bartaburu, D., Duarte, E., Montes, E., Pereira, M., Corral, J., Lucena, C. J., & Morales, G. H. (2014). A further step towards participatory modelling. Fostering stakeholder involvement in designing models by using executable UML. *Journal of Artificial Societies and Social Simulation, 17*, 6.

Bousquet, F. (1994). *Des milieux, des poissons, des hommes: Étude par simulations multi-agents. Le cas de la pêche dans le delta central du Niger.* PhD. in Economics. Lyon: Université Lyon 1.

Bousquet, F., Le Page, C., Bakam, I., & Takforyan, A. (2000). A spatially-explicit individual-based model of blue duikers' population dynamics: Multi-agent simulations of bushmeat hunting in an eastern Cameroonian village. *Ecological Modelling, 138*, 331–346.

Bonaudo, T. (2005). *La gestion environnementale sur un front pionnier amazonien.* PhD. in Agronomy,. Paris: Institut National Agronomique Paris-Grignon.

Bonnefoy, J.-L., Bousquet, F., & Rouchier, J. (2001). Modélisation d'une interaction individus, espace, société par les systèmes multi-agents: Pâture en forêt virtuelle. *L'espace géographique, 1*, 13–25.

Bouleau, N. (2001). *Modélisation et contre-expertise* (pp. 462–468). Paris: APMEP.

Brenner, T., & Werker, C. (2007). *A practical guide to inference in simulation models, papers on economics and evolution.* Jena: Max Planck Institute of Economics Evolutionary Economics Group.

Bonaparte, Napoléon. Declaration before the battle of the Pyramids. Cairo, 21/07/1798.

Carley, K. M. (1996). *Validating computational models.* Pittsburgh: Carnegie Mellon University.

Carrer, F. (2013). An ethnoarchaeological inductive model for predicting archaeological site location: A case-study of pastoral settlement patterns in the Val di Fiemme and Val di Sole (Trentino, Italian Alps). *Journal of Anthropological Archaeology, 32*, 54–62.

Carozza, L., Berger, J. F., Burens-Carozza, A., & Marcigny, C. (2015). Society and environment in Southern France from the 3rd millennium BC to the beginning of the 2nd millennium BC: 2200 BC a tipping point?. 2200BC-A climatic breakdown as a cause for the collapse of the old world? *Tagungen des landesmuseums für vorgeschichte halle, Band, 12*, 333–362.

Castella, J.-C., Boisseau, S., Trung, T. N., & Quang, D. D. (2005). Agrarian transition and lowland-upland interactions in mountain areas in northern Vietnam: Application of a multi-agent simulation model. *Agricultural Systems, 86*(3), 312–332.

Castella, J.-C., Suan, P. K., Quang, D. D., Verburg, P. H., & Thai, H. C. (2007). Combining top-down and bottom-up modelling approaches of land use/cover change to support public policies: Application to sustainable management of natural resources in northern Vietnam. *Land Use Policy, 24*(3), 531–545.

Castella, J.-C., & Verburg, P. H. (2007). Combination of process-oriented and pattern-oriented models of land-use change in a mountain area of Vietnam. *Ecological Modelling, 202*(3–4), 410–420.

Chang-Martínez, L. A., Mas, J.-F., Valle, N. T., & Torres, P. S. U. (2015). Modeling historical land cover and land use: A review from contemporary modeling. *ISPRS International Journal of Geo-information, 4*, 1791–1812. https://doi.org/10.3390/ijgi4041791.

Chattoe, E. (2000). *Why is building Multi-Agent Models of social systems so difficult? A case study of innovation diffusion.* 24th International Conference of Agricultural Economists (IAAE), Berlin, Germany.

Cioffi-Revilla, C., & Rouleau, M. (2009). *MASON RebeLand: An Agent-Based Model of Politics, Environment, and Insurgency!* Annual Conference of the European Social Sciences Association, Brescia, Italy.

Conolly, J., Colledge, S., & Shennan, S. (2008). Founder effect, drift, and adaptive change in domestic crop use in early Neolithic Europe. *Journal of Archaeological Science, 35,* 2797–2804. https://doi.org/10.1016/j.jas.2008.05.006.

Couclelis, H. (2002). *Modeling frameworks, paradigms, and approaches. Geographic information systems and environmental modelling.* London: Prentice Hall.

Courdier, R., Marcenac, P., & Giroux, S. (1998). Un processus de développement en spirale pour la simulation multi-agents. *L'Objet, 4,* 73–86.

Deguet, J., Demazeau, Y., & Magnin, L. (2006). Elements about the emergence issue: A survey of emergence definitions. *Complexus, 3*(1–3), 24–31.

Dolukhanov, P., & Shukurov, A. (2003). Modelling the Neolithic dispersal in northern Eurasia. *Documenta Praehistorica, 31,* 35–47.

Ducrot, R., & Botta, A. (2009). *Strategies to institutionalize companion modelling approaches.* 18th World IMACS/MODSIM Congress. 2983–2990.

Ebersbach, R. (1999). Modeling Neolithic agriculture and stock-farming at Swiss Lake shore settlements:-evidence from historical and ethnographical data. *Archaeofauna: International Journal of Archaeozoology, 8,* 115–122.

Ebersbach, R. (2004). Agriculture, stock farming and environment: Adaptation and change during the Neolithic lakeshore period (4300-2400 cal BC) in Switzerland. *Antaeus, 27,* 287–292.

Ebersbach, R., & Schade, C. (2004). Modelling the intensity of linear pottery land use: An example from the Märlener Bucht in the Wetterau Basin, Hesse, Germany. In R. Ebersbach & C. Schade (Eds.), *Enter the past: The E-way into the four dimensions of cultural heritage* (pp. 337–348). Oxford: BAR International Series.

Eco, U. (1990). *The limits of interpretation.* New Haven: Indiana University Press.

Edmonds, B., & Moss, S. (2005). From KISS to KIDS: An "anti-simplistic" modelling approach. *Lecture Notes in Artificial Intelligence, 34,* 130–144.

Epstein, J. M. (1999). Agent-based computational models and generative social science. *Complexity, 4,* 41–60.

Etienne M. (2010). La modélisation d'accompagnement: Une démarche participative en appui au développement durable. Editions QUAE, Montpellier, France.

Etienne, M., DuToit, D., & Pollard, S. (2011). ARDI: A co-construction method for participatory modelling in natural resources management. *Ecology & Society, 16,* 44.

Epstein, J. M. (2008). Why Model? *Journal of Artificial Societies and Social Simulation, 11*(4), 12.

Franck, U., & Troitzsch, K. G. (2005). Epistemological Perspectives on Simulation. *Journal of Artificial Societies and Social Simulation, 8*(4), 7.

Georgeff, M., Pell, B., Pollack, M., Tambe, M., & Wooldridge, M. (1999). The belief-desire-intention model of agency. In J. Müller, A. Rao, & M. Singh (Eds.), *Intelligent agents vs. Agents theories, architectures and languages* (pp. 1–10). Berlin Heidelberg: Springer.

Gibon, A., Sheeren, D., Monteil, C., Ladet, S., & Balent, G. (2009). Modelling and simulating change in reforesting mountain landscapes using a social-ecological framework. *Landscape Ecology, 25*(2), 267–285.

Gilbert, N., & Troitzsch, K. (2005). *Simulation for the social scientist.* New York: McGraw-Hill Education.

Graves, D. (2011). The use of predictive modelling to target Neolithic settlement and occupation activity in mainland Scotland. *Journal of Archaeological Science, 38,* 633–656. https://doi.org/10.1016/j.jas.2010.10.016.

Gregg, S. A. (1988). Foragers and farmers: Population interaction and agricultural expansion in prehistoric Europe. In *Prehistoric archaeology and ecology series*. Chicago: University of Chicago Press.

Grimm, V., Berger, U., Bastiansen, F., Eliassen, S., Ginot, V., Giske, J., Goss-Custard, J., Grand, T., Heinz, S. K., Huse, G., Huth, A., Jepsen, J. U., Jørgensen, C., Mooij, W. M., Müller, B., Pe'er, G., Piou, C., Railsback, S. F., Robbins, A. M., Robbins, M. M., Rossmanith, E., Rüger, N., Strand, E., Souissi, S., Stillman, R. A., Vabø, R., Visser, U., & De Angelis, D. L. (2006). A standard protocol for describing individual-based and agent-based models. *Ecological Modelling, 198*, 115–126. https://doi.org/10.1016/j.ecolmodel.2006.04.023.

Grimm, V., Berger, U., DeAngelis, D. L., Polhill, J. G., Giske, J., & Railsback, S. F. (2010). The ODD protocol: A review and first update. *Ecological Modelling, 221*, 2760–2768. https://doi.org/10.1016/j.ecolmodel.2010.08.019.

Hazelwood, L., & Steele, J. (2004). Spatial dynamics of human dispersals: Constraints on modelling and archaeological validation. *Journal of Archaeological Science, 31*, 669–679. https://doi.org/10.1016/j.jas.2003.11.009.

Henrickson, L., & McKelvey, B. (2002). Foundations of "new" social science: Institutional legitimacy from philosophy, complexity science, postmodernism, and agent-based modeling. *Proceedings of the National Academy of Sciences, 99*, 7288–7295. https://doi.org/10.1073/pnas.092079799.

Henry, S., Boyle, P., & Lambin, E. F. (2003). Modelling inter-provincial migration in Burkina Faso, West Africa: The role of sociodemographic and environmental factors. *Applied Geography, 23*, 115–136.

Janssen, M. A., & Ostrom, E. (2006a). Empirically based, agent-based models. *Ecology and Society, 11*, 24–37.

Janssen, M. A., & Ostrom, E. (2006b). Governing Social-Ecological Systems. In Handbook of Computational Economics, édité par L. Tesfatsion et K.L. Judd, 2:1465-1509, Chap. 30. Elsevier. https://doi.org/10.1016/S1574-0021(05)02030-7.

Janssen, M. A. (2009). Understanding artificial anasazi. *Journal of Artificial Societies and Social Simulation, 12*, 13.

Janssen, M. A., Kohler, T. A., & Scheffer, M. (2003). Sunk-cost effects and vulnerability to collapse in ancient societies. *Current Anthropology, 44*, 722–728.

Janssen, M. A., & Scheffer, M. (2004). Overexploitation of renewable resources by ancient societies and the role of sunk cost effects. *Ecology & Society, 9*, 6.

Jonker, C. M., & Treur, J. (2013). A formal approach to building compositional agent-based simulations. In B. Edmonds & R. Meyer (Eds.), *Simulating social complexity: A handbook* (pp. 57–94). Berlin, Heidelberg: Springer. https://doi.org/10.1007/978-3-540-93813-2_5.

Kaplan, J. O., Krumhardt, K. M., Pfeiffer, M., Davis, B. A. S., & Zanon, M. (2012). *From forest to farmland and meadow to metropolis: Integrated modeling of Holocene land cover change*. 97th ESA Annual Meeting, Portland, USA.

Kieken, H., Dahan, A., & Armatte, M. (2003). Models and modelling processes: A critical step for an environmental research. *Natures Sciences Sociétés, 11*, 396–403.

Kohler, T. A., & Gumerman, G. J. (2000). *Dynamics in human and primate societies: Agent-based modeling of social and spatial processes*. Oxford: Oxford University Press.

Kohler, T. A., & van der Leeuw, S. E. (2007). *The model-based archaeology of socionatural systems*. Oxford: Oxbow Books Ltd.

Kohler, T. A., Bocinsky, K. R., Cockburn, D., Crabtree, S. A., Varien, M. D., Kolm, K. E., Smith, S., Ortman, S. G., & Kobti, Z. (2012). Modelling prehispanic Pueblo societies in their ecosystems. *Ecological Modelling, 241*, 30–41. https://doi.org/10.1016/j.ecolmodel.2012.01.002.

Kreuz, A., Marinova, E., Schäfer, E., & Wiethold, J. (2005). A comparison of early Neolithic crop and weed assemblages from the Linearbandkeramik and the Bulgarian Neolithic cultures: Differences and similarities. *Vegetation History and Archaeobotany, 14*, 237–258.

Lake, M. W. (2015). Explaining the past with ABM: On modelling philosophy. In G. Wurzer, K. Kowarik, & H. Reschreiter (Eds.), *Agent-based modeling and simulation in archaeology. Advances in geographic information science*. Cham: Springer.

Lambin, E. F., Rounsevell, D. A. M., & Geist, H. J. (2000). Are agricultural land-use models able to predict changes in land-use intensity? *Agriculture, Ecosystems & Environment, 82*(1–3), 321–331.

Lambin, E. F., Turner, B. L., Geist, H. J., Agbola, S. B., Angelsen, A., Bruce, J. W., Coomes, O. T., Dirzo, R., Fischer, G., Folke, C., Goerge, P. S., Homewood, K., Imbernon, J., Leemans, R., Xiubin, L., Moran, E. F., Mortimore, M. J., Ramakrishnan, P. S., Richards, J. F., Skanes, H., Steffen, W., Stone, G. D., Svedin, U., Veldkamp, T. A., Vogel, C., & Jianchu, X. (2001). The causes of land-use and land-cover change: Moving beyond the myths. *Global Environmental Change, 11*, 261–269.

Landais, É. (1992). Principes de modélisation des systèmes d'élevage. Approches graphiques. *Les Cahiers de la recherche-développement, 32*, 82–95.

Lemmen, C., Wirtz, K. W., & Gronenborn, D. (2009). *Prehistoric land use and Neolithisation in Europe in the context of regional climate events*. EGU General Assembly 2009.

Lemmen, C., & Khan, A. (2012). A simulation of the Neolithic transition in the Indus valley. Chapter manuscript AGU monograph "Climates, landscapes and civilizations" arXiv preprint arXiv:1110.1091; Cornell University, https://arxiv.org/abs/1110.1091.

Lemmen, C., & Wirtz, K. W. (2014). On the sensitivity of the simulated European Neolithic transition to climate extremes. *Journal of Archaeological Science, 51*, 65–72. https://doi.org/10.1016/j.jas.2012.10.023.

Le Page, C., Bazile, D., Becu, N., Bommel, P., Bousquet, F., Etienne, M., Mathevet, R., Souchère, V., Trébuil, G., & Weber, J. (2013). Agent-based modelling and simulation applied to environmental management: A review. In B. Edmonds & R. Meyer (Eds.), *Simulating social complexity: A handbook* (pp. 499–540). Berlin Heidelberg: Springer.

Lespez, L., Carozza, L., Berger, J. -F., Kuzucuoglu, C., Ghilardi, M., Carozza, J. -M., Vannière, B., & The ArcheoMed Team. (2016). Rapid climatic change and social transformations: Uncertainties, adaptability and resilience. In S. Thiébault & M. Jean-Paul (Eds.), *The Mediterranean region under climate change: A scientific update* (pp. 35–45). Marseille: IRD; AllEnvi. ISBN 978-2-7099-2219-7.

Lieurain, E. (1998). *Couplage SIG-SMA-SGBD*. Montpellier: CIRAD-Tera.

Livet, P., Phan, D., & Sanders, L. (2014). Diversité et complémentarité des modèles multi-agents en sciences sociales. *Revue française de sociologie, 55*, 689–729. https://doi.org/10.3917/rfs.554.0689.

Malawska, A., & Topping, C. J. (2016). Evaluating the role of behavioral factors and practical constraints in the performance of an agent-based model of farmer decision making. *Agricultural Systems, 143*, 136–146. https://doi.org/10.1016/j.agsy.2015.12.014.

Malézieux, E., Crozat, Y., Dupraz, C., Laurans, M., Makowski, D., Ozier-Lafontaine, H., Rapidel, B., de Tourdonnet, S., & Valantin-Morison, M. (2009). Mixing plant species in cropping systems: Concepts, tools and models. A review. *Agronomy for Sustainable Development, 29*, 43–62.

Maru, Y. T., Alexandridis, K., & Perez, P. (2009). *Taking "participatory" in participatory modelling seriously*. 18th World IMACS/MODSIM Congress, 3011–3017.

Mathevet, R., Bousquet, F., Le Page, C., & Antona, M. (2003). Agent-based simulations of interactions between duck population, farming decisions and leasing of hunting rights in the Camargue. *Ecological Modelling, 165*, 107–126.

Maury, M. (2004). *Modélisation de la divagation des troupeaux de bovins (M.Sc. Cognitive Sciences)*. Bordeaux: Université Bordeaux II.

Mazher, A. K. (2001). *Towards a unified approach to modeling and computer simulation of social systems. Part I: Methodology of model construction*. Paris: Seuil.

Mazoyer, M., & Roudart, L. (1997). *Histoire des agricultures du monde*. Paris: Seuil.

Mechoud, S., Hill, D. R. C., Campos A., Orth, D., Michelin, Y., Poix, C., L'Homme, G., et al. (2000). Simulation Multi-Agents de l'entretien du paysage par des herbivores en moyenne montagne. In SMAGET Proceedings, 101–20. Clermont-Ferrand, France: CEMAGREF Editions. http://wwwlisc.clermont.cemagref.fr/Animation/SeminairesColloquesRealises/smaget/Interventions/mechoud/ArticleTernant.html

Montmain, J., & Penalva, J.-M. (2003). *Choix publics stratégiques et systèmes sociaux: Etat de l'art sur les théories de la décision et méthodologies de l'approche système*. Paris: LGI2P.

Minsky, M. (1965). Matter, mind and models. *Proceedings of the International Federation of Information Processing Congress 1965, 1*, 45–49.

Morris, M. D. (1991). Factorial sampling plans for preliminary computational experiments. *Technometrics, 33*, 161–174.

Munroe, D. K., & Müller, D. (2007). Issues in spatially explicit statistical land-use/cover change (LUCC) models: Examples from western Honduras and the Central Highlands of Vietnam. *Land Use Policy, 24*, 521–530.

Neuhauser, C., & Pacala, S. W. (1999). An explicitly spatial version of the Lotka-Volterra model with interspecific competition. *The Annals of Applied Probability, 9*, 1226–1259.

Nachi, M. (2006). Concept commun et concept analogique de compromis: Un air de famille. Essai d'épistémologie pragmatique. *SociologieS, Théories et recherches*. http://sociologies.revues.org/3097.

Ostrom, T. M. (1988). Computer simulation: The third symbol system. *Journal of Experimental Social Psychology, 24*(5), 381–392. https://doi.org/10.1016/0022-1031(88)90027-3.

Paegelow, M., & Camacho Olmedo, M. T. (2008). *Modelling environmental dynamics: Advances in geomatic solutions*. Berlin, Heidelberg: Springer.

Parker, D. C., Berger, T., & Manson, S. M. (2001). *Agent-based models of land-use and land-cover change: Report and review of in international workshop* (Vol. LUCC Reports, 6). Irvine: UC Irvine.

Patterson, M. A., Sarson, G. R., Sarson, H. C., & Shukurov, A. (2010). Modelling the Neolithic transition in a heterogeneous environment. *Journal of Archaeological Science, 37*, 2929–2937.

Pipino, L. L., Lee, Y. W., & Wang, R. Y. (2002). Data quality assessment. *Communications of the ACM, 45*(4ve), 211–218.

Popper, K. R. (1985). *Conjectures et réfutations. La croissance du savoir scientifique*. Paris: Payot.

Porter, A. L., Roessner, J. D., Cohen, A. S., & Perreault, M. (2006). Interdisciplinary research: Meaning, metrics and nurture. *Research Evaluation, 15*, 187–195. https://doi.org/10.3152/147154406781775841.

Premo, L. S. (2006). Exploratory agent-based models: Towards an experimental ethnoarchaeology. In *Digital discovery: Exploring new frontiers in human heritage. CAA* (pp. 29–36). Budapest: Archeolingua Press.

Radja, K. (2003). *La famille dans l'analyse économique: Modélisation et représentations théoriques de la famille*. Bordeaux: Université Montesquieu Bordeaux IV.

Rodgers, J. L. (2016). Moving in parallel toward a modern modeling epistemology: Bayes factors and frequentist modeling methods. *Multivariate Behavioral Research, 51*, 30–34. https://doi.org/10.1080/00273171.2015.1093459.

Rösch, M., Ehrmann, O., Hermann, L., Schulz, E., Bogenrieder, A., Goldammer, J. P., Hall, M., Page, H., & Schier, W. (2002). An experimental approach to Neolithic shifting cultivation. *Vegetation History and Archaeobotany, 11*(1–2), 143–154.

Rouchier, J., Bousquet, F., Barreteau, O., Le Page, C., & Bonnefoy, e. J.-L. (1979). Multi-agent modelling and renewable resources issues: the relevance of shared representations for interacting agents. *Lecture Notes in Artificial Intelligence* (181), 197.

Rouchier, J., Bousquet, F., Requier-Desjardins, M., & Antona, M. (2001). A multi-agent model for describing transhumance in North Cameroon: Comparison of different rationality to develop a routine. *Journal of Economic Dynamics and Control, 25*, 527–559.

Rowley-Conwy, P. A. (1981). Slash and burn in the temperate European Neolithic. In R. J. Mercer (Ed.), *Farming practice in british prehistory* (pp. 85–96). Edinburgh: Edinburgh University Press.

Rubio-Campillo, X. (2015). Large simulations and small societies: High performance computing for archaeological simulations. In G. Wurzer, K. Kowarik, & H. Reschreiter (Eds.), *Agent-based modeling and simulation in archaeology. Advances in geographic information science*. Vienna: Springer.

Rumbaugh, J., Jacobson, I., & Booch, G. (2004). *The unified modeling language reference manual* (2nd ed.). London: Pearson plc.

Saqalli, M., Bielders, C. L., Defourny, P., & Gérard, B. (2010a). Simulating rural environmentally and socio-economically constrained multi-activity and multi-decision societies in a low-data context: A challenge through empirical agent-based modeling. *Journal of Artificial Societies and Social Simulation, 13*, 1–20.

Saqalli, M., Gérard, B., Bielders, C. L., & Defourny, P. (2010b). Testing the impact of social forces on the evolution of Sahelian farming systems: A combined agent-based modeling and anthropological approach. *Ecological Modelling, 221*, 2714–2727.

Saqalli, M., Bielders, C. L., Defourny, P., & Gérard, B. (2013). Reconstituting family transitions of Sahelian western Niger 1950–2000: An agent-based modelling approach in a low data context. *CyberGeo, 634*. https://doi.org/10.4000/cybergeo.25760.

Saqalli, M., Salavert, A., Bréhard, S., Bendrey, R., Vigne, J.-D., & Tresset, A. (2014). Revisiting and modelling the woodland farming system of the early Neolithic Linear Pottery Culture (LBK), 5600-4900 B.C. *Vegetation History and Archaeobotany, 23*, 37–50. https://doi.org/10.1007/s00334-014-0436-4.

Saqalli, M., & Baum, T. G. (2016). Pathways for scale and discipline reconciliation: Current socio-ecological modelling methodologies to explore and reconstitute human prehistoric dynamics. In J. A. Barceló & F. del Castillo (Eds.), *Simulating prehistoric and ancient worlds* (pp. 233–255). Springer Computational Social Sciences. Springer.

Saqalli M., Maestripieri N., Jourdren M., Saenz M., & Maire E. (2018a). Spatialiser un risque environnemental via les perceptions locales : Une démarche, trois terrains (Equateur, Tunisie, Laos). In: M. Gaille. Pathologies environnementale: Identifier, comprendre, agir. Chap. 2 77-112. CNRS Editions. Paris, France.

Saqalli, M., Chakroun, H., & Mahé, G. (2018b). Légitimité des scénarios de gouvernance et métrique de soutenabilité. Rapport SICMED, Marrakech conference, Morocco.

Schulze, J., Müller, B., Groeneveld, J., & Grimm, V. (2017). Agent-based modelling of social-ecological systems: Achievements, challenges, and a way forward. *Journal of Artificial Societies and Social Simulation, 20*, 8.

Schutte, S. (2010). Optimization and falsification in empirical agent-based models. *Journal of Artificial Societies and Social Simulation, 13*, 2.

Soulié, J.-C., & Thébaud, O. (2006). Modelling fleet response in regulated fishery: An agent-based approach. *Mathematical and Computer Modelling, 44*(5–6), 553–564.

Taber, C. S., & Timpone, R. J. (1996). *Computational modeling. Quantitative applications in the social sciences. 113*. Thousand Oaks, London: Sage.

Tipping, R., Bunting, M. J., Davies, A. L., Murray, H., Fraser, S., & McCulloch, R. (2009). Modelling land use around an early Neolithic timber "hall" in north east Scotland from high spatial resolution pollen analyses. *Journal of Archaeological Science, 36*, 140–149. https://doi.org/10.1016/j.jas.2008.07.016.

Todd, E. (2011). *L'origine des systèmes familiaux Tome 1: L'Eurasie*. Paris: Gallimard.

Turner, B. L. (2002). Contested identities: Human-environment geography and disciplinary implications in a restructuring academy. *Annals of the Association of American Geographers, 92*, 52–74.

Van Gigch, J. P. (1993). Metamodeling: The epistemology of system science. *Systems Practice, 6*, 251–258.

Van Vliet, N., Milner-Gulland, E. J., Bousquet, F., Saqalli, M., & Nasi, R. (2010). Effect of small-scale heterogeneity of prey and hunter distributions on the sustainability of bushmeat hunting. *Conservation Biology, 24*, 1327–1337.

Verburg, P. H., Schot, P. P., Dijst, M. J., & Veldkamp, A. (2004). Land use change modelling: Current practice and research priorities. *GeoJournal, 61*(4), 309–324.

Verhagen, P., & Whitley, T. G. (2012). Integrating archaeological theory and predictive modeling: A live report from the scene. *Journal of Archaeological Method and Theory, 19*, 49–100.

Wirtz, K. W., & Lemmen, C. (2003). A global dynamic model for the Neolithic transition. *Climatic Change, 59*, 333–367.

Wolfram, S. (2002). *A new kind of science*. Champaign: Wolfram Media.
Youngblood, D. (2007). Interdisciplinary studies and the bridging disciplines: A matter of process. *Journal of Research Practice, 3*, M18.
Ye, M., & Hill, M. C. (2017). Global sensitivity analysis for uncertain parameters, models and scenarios. In G. Petropoulos & P. K. Srivastava (Eds.), *Sensitivity analysis in earth observation modelling* (pp. 177–210). Amsterdam: Elsevier.
Yu, Y., Guo, Z., Wu, H., & Finke, P. A. (2012). Reconstructing prehistoric land use change from archaeological data: Validation and application of a new model in Yiluo valley, northern China. *Agriculture, Ecosystems & Environment, 156*, 99–107. https://doi.org/10.1016/j.agee.2012.05.013.

From Culture Difference to a Measure of Ethnogenesis: The Limits of Archaeological Inquiry

Juan A. Barceló, Florencia Del Castillo, Laura Mameli, Franceso J. Miguel, and Xavier Vilà

Introduction

The Idea of Ethnicity

From some years now, we have been working in the problem of identifying social identity in the archaeological record. Beyond the mere identification of "groups" or "cultures," our research has been focused on the social mechanics of identity formation, negotiation, and imposition in societies for which we do not have any textual data, but only some archaeological observations. Because "culture" or "ethnicity" are not directly observable categories, the debate about ethnicity is one of the most controversial issues in archaeology (Clarke 1968; Emberling 1997; Shennan 1989; Dietler 1994; Jones 1997; Demoule 1999; Meskell 2002; Hirschmann 2004; O'Brien and Lyman 2004; Lucy 2005; Casella and Fowler 2005; Hudson 2006; Fernández Götz 2008; Voss 2008; Watkins 2008; Bellon et al. 2009; García Fernández and Bellon 2009; Hales and Hodos 2010; Mayor 2010; Hu 2013; Kovacevic et al. 2015; Sommer 2016). Among the different identities an individual can build about herself/himself, if there is an "ethnical" identity, it should be related with some kind of "collective identity", that is, what a number of social agents believe they share with other social agents belonging to the same group *as a consequence of this belief*. It is then an attribute of the group, and not only a feature of the individual. The only way to deal with this complex issue is by making emphasis on

---

J. A. Barceló (✉) · L. Mameli · F. J. Miguel · X. Vilà
Universitat Autònoma de Barcelona, Barcelona, Spain
e-mail: juanantonio.barcelo@uab.cat

F. Del Castillo
CONICET- Centro Nacional Patagónico (CCT-CENPAT-Argentina),
Puerto Madryn, Argentina

© Springer Nature Switzerland AG 2019
M. Saqalli, M. Vander Linden (eds.), *Integrating Qualitative and Social Science Factors in Archaeological Modelling*, Computational Social Sciences,
https://doi.org/10.1007/978-3-030-12723-7_3

the social mechanisms by which a social agent is considered – by itself or by others – as a member of some group (Barth 1969; Bentley 1987).

A feeling of belonging to a differentiated group emerges from the usual preference for interacting with others who share similar traits and practices. Individuals may display "in-group favoritism" (Hammond and Axelrod 2006; Efferson et al. 2008) also called "parochialism" (Bowles and Gintis 2004; Koopmans and Rebers 2009) in choosing how to interact, based on the advantages they win when interacting with "others" (according to individual or global beliefs). Subjects have high expectations of the contributions of in-group members. As a consequence, their own behavior will be strongly conditioned on other group members' expected behavior (Koopmans and Rebers 2009). Thomas Schelling proved many years ago how this social mechanism generates a partition in the population in such a way that social clusters emerge (Schelling 1971; Clark and Fossett 2008; Aldén et al. 2015). In other words, identity-constrained cooperation and in-group bias are, to some extent, linked.

In some sense, ethnicity may be defined as the conscious maximization of in-group homogeneity and within-groups heterogeneity, in such a way that social agents tend to approach other social agents through a self-reinforcing mechanism of "more interaction then more similarity" (Axelrod 1997). Communality of action gradually emerges over time, and that communality will affect some other behaviors or features, like biological phenotype, language, or material culture. What is important is not so much the group members' awareness of a supposed common origin but the repetition over time of the same interactions with the same people in the same sense, that is, the *history* of interactions. Only when such a pattern of visible similarities and differences is maintained from generation to generation – it constitutes an historical trajectory – the resulting identity could be labelled or recognized as ethnical.

In this way, the higher the cultural similarity between members of the same group, the more social constraints on reproduction mechanisms, and then a higher level of genetic similarity among descendants will be expected. The least cultural similarity in a population, the less social constrains on reproduction, and the more mixed their descendants, who will not share any common genetic background. In other words, the similarity in biological phenotype among members of an "ethnic" group is not what defines the group, but the result of the way agents interact socially to choose a reproductive mate within an already defined group (Abruzzi 1982; Whitmeyer 1997). Human reproduction is not just a mere biological process but a socially mediated mechanism. Reproductive mates are consciously chosen, and many social, ideological, and political constraints impose some directionality in social reproduction (Bernardi 2003; Bongaarts and Watkins 1996; Kalick and Hamilton 1986).

A group of people adopts a common identity because its social activity and social reproduction mechanisms have persisted through a certain number of changes (Bate 1998: 95). The concrete way ethnicity is expressed may vary from group to group, from historical circumstance to historical circumstance. What defines the group (its "identity" at high scale) is usually a consequence of what members of such a group have learnt from their antecessors, and what they have learnt is a consequence of what those antecessors consciously decided to transmit to their descendants. Long-term communalities resulting from the repetition of interactions through generations is what constitutes the "ethnical" aspect of collective identity. In our view, *ethnic identity is the result of a social mechanism of production and reproduction of intentionally produced similarities and differences in belief and behavior, which are used by a particular group of people to express their own internal coherence and explicit differentiation from neighboring social groups.* On the other hand, we assume that communalities in belief and action do not exist forever because social similarity is in the process of continuous building, influenced by the very many aspects of social life. They are learned and shared across people. The challenge to this view is that instead of *assuming* that agents have common identity traits based on membership to an already existing "ethnic" group, agents ask themselves about the extent to which they are similar or different to others in the neighborhood (Romney et al. 1986, 1996; Romney 1999; Garro 2000; Weller 2007; Sieck 2010).

It is in this sense that *ethnicity has an indissoluble historical component* (Alba 1990): similarity and regularity are the obvious consequence of maintaining the same mechanisms of constrained interaction and cooperation for quite a long time. People with the same genealogical trajectory will show a degree of similarity in their motivations, goals, actions, behaviors, and mediating artifacts which do not depend on their actual will but on what they have received from the past (Dow et al. 1984; Eff 2004). The more inter-generational knowledge transmission among socially aggregated individuals in the past, the more similar is the social activity performed by the agents in the present and also their actions and the material and immaterial consequences of their actions in the future. In other words, the present material effects of "ethnicity" – as an action coordination identity bias – could be explained as an aggregate outcome of past coordination practices. So, we move from the domain of non-observable beliefs to the domain of observational records.

Sean Jones (1997) has argued that ethnic identity in the past is beyond the reach of archaeology, as the meaning initially attached to the material symbols used for the construction of ethnic boundaries will always be unknown. In a typically empiricist stance, Brather (2004) recommends that archaeologists abandon any research into ethnicity, as long as there are no ethnographic sources or independent historical texts to decipher the meaning of these symbols. If this were so, it would not only be a drama for archaeology, condemned to be a discipline limited to the description of ancient artifacts, but it would also make it totally impossible to study ethnicity as a historical process, since it would mean the non-observability of ethnogenesis. The differences observed in the present would have no explanation because there would be no way to observe their formation in the past. Although ethnicity should be considered as a symbolic rather than a material construct, and therefore not necessarily being claimed to exist in a material way, it has affected the material world in a tangible way, and that is what we must investigate. If the determination of ethnicity depended solely on the subjectivity of the agents and was totally situational, nonnormative, and eternally changing, nothing could really be defined as ethnic. If the identification that the members of a group feel is a mere result of individual subjectivity, without a minimum objective consensus about what really caused that identification, ethnicity becomes a useless categorization, without the slightest heuristic value.

We cannot accept a vision of ethnicity that results only from what people believe (or say they believe) about their identity and the ideas they have about their origin. To affirm this would mean giving a charter of nature to all forms of ethnic cleansing expressed by political interests and that perverts the way a human community has been historically constructed (Bogdanovic 2011).

The error is in the cognitive sense of the term "culture." A theory of culture in terms of cognitive elements and structures currently dominates the academic realm of anthropology and the social sciences. Culture appears as the means through which people transform the material world into an environment of symbols to which they give meaning and value (Geertz 1973; Bentley 1987; Cosgrove and Jackson 1987; Schudson 1989; Handwerker 2008; Fischer 2007). The consequence to which such a conception leads us is an absurd imprecision when defining the concept. According to Barth (1969), the very lack of precision of the definition of "culture" is precisely what has led to the abuse of the term and of all related to it. It has allowed anthropologists and archaeologists to essentialize social dynamics through typologies, shifting interest from empirical observation analysis to simplification in abstract models without sufficient explanatory power. In rephrasing an abstraction, the concepts of "culture" and "ethnicity" reinforce our subjective assumptions, rather than allowing us to confront them with what we can get to know from the real action of real people. At the same time, imprecise abstraction denies the existence of real noncultural (i.e., physical) aspects that may have affected, conditioned, and/or determined the human action and activity of individuals.

Culture as a Measurement

Identity (also called *sameness*) is whatever makes a social agent definable and recognizable, in terms of possessing a set of qualities or characteristics that distinguish she/he from agents of a different group (Williams 1989; Deutsch 1997; Noonan and Curtis 2017). Or, in layman's terms, identity is whatever makes someone "similar" to a particular group of social agents. Two social agents will be similar if, at corresponding moments of time, and at corresponding points in space, they believed the same, and consequently they did the same, with a similar material consequence.

Perceiving what makes my neighbor similar or different to me is not always a strictly rational operation. Then, how I know I am a member of a particular group? In the previous section, we have rejected the usual view of a single cultural trait or "meme" being necessary for membership in an identity group. We have stressed the fact that *any* cultural trait can be sufficient for defining identity. This assumption is of relevance for archaeological studies, because it clearly indicates that ethnical identity does not reside in a single category of material objects (Clarke 1968; Hodder 1982; McGuire 1982; Sackett 1990; Braun 1991; Longacre and Stark 1992;

Larick 1991; Thomas 1996; Emberling 1997). Does it preclude the study of ethnicity in archaeology?

When a social agent "believes" that it shares something with neighboring social agents, it adjusts its behavior to increase the *regularity* within this group of social agents, what means that certain social activities will be more *probable* in that group than other kind of activities, which would not be shared. Following the most habitual definition of probability, an aggregate of social agents would exhibit some degree of regularity when a comparatively high number of social agents are involved in the most "frequent" social activities, and low numbers of social agents are involved in "infrequent" activities. Increasing regularity in social behavior has the advantage of increasing useful redundancy into the mechanism. As an obvious result, it emerges a similarity pattern in the particular distribution of information (ideas, beliefs, concepts, symbols, technical knowledge, etc.) across this population as a result of the repetition of the same activities between the same people. This pattern of similarity is what we may call *culture* (D'Andrade 1987; Carley 1991; Mosterín 1993; Axelrod 1997; Boyd and Richerson 2005; Zou et al. 2009; Squazzoni 2012), and it is the consequence of the sense of community generated by the restriction of interaction with agents considered to be similar. What really defines "culture" and "ethnic identity" is the "relational" character of that set of properties or attributes, so that no part can be changed without affecting the other parts (Peroff 1997; Fischer 2007).

Therefore, an aggregation of similar activities and social practices repeated at the same place over a period of time by the same group of people is the best estimate for communality in belief and behavior and, hence, an evidence for *culture* and shared *identity*. This is what Romney has called "cultural consensus" (Romney and Weller 1984; Romney et al. 1986; Romney 1999). Its proper definition rests either on exact or approximate repetitions of social activities by the same agents at different moments of time. The accuracy of that aggregation depends on the agreement between what people did and thought and the number of observations on past actions and believes (Romney 1999; Romney and Weller 1984; Romney et al. 1986; Garro 2000; Weller 2007; Sieck 2010; Dressler et al. 2005; Borgatti and Halgin 2011).

We think that differences in "culture" express the expected variance in a distribution of social activities among synchronous human aggregates or populations. Obviously, simple agreement about a set of items does not imply that the population may have any degree of identity. The extent to which social activities are actually shared in any given population is an empirical question. The analysis of communalities does not create communalities; it only measures the degree some consensus may appear at different circumstances. What we need is a measure of "social similarity," to be analyzed quantitatively as the consequence of the particular way those social agents have interacted, aggregated in space and time because of some of these interactions, and reproduced the basis of such an aggregation. The strength of the feeling of cultural similarity depends on how many of the characteristics individuals believe they have in common with others in the group. Obviously, different communities at different moments will attain different degrees of social similarity. On some circumstances, the attained degree compared with the threshold of a particular utility function and some particular new behavior will be adopted consequently. Such threshold is also variable according to the social environment and the circumstances at that precise moment of time.

In a 1977 paper, Amos Tversky proposed his feature contrast model of similarity. It can be used as a way to build "cultural" similarity indexes in the form of mathematical distances between different social aggregates. Such a feature set is the set of logic predicates, which are true for the population in question. Let a and b be two aggregates of social agents; A and B are the respective sets of features, and $s(a, b)$ represents a measure of the similarity between a and b.

Using Tversky's approach, the similarity of social aggregate a to social aggregate b is a function of (1) the information traits ("culture") common to a and b ("A and B"), (2) those in a but not in b (symbolized "A − B"), and (3) those in b but not in a ("B − A"). Note especially that similarity is not just a function of common features but depends also on features that are unique to each population and that their relative importance varies with the parameters x, y, and z.

$$S_{a,b} = xf(A \text{ and } B) - yf(A - B) - zf(B - A)$$

Here $S_{a,b}$ is an interval scale of similarity, f is a parameter that reflects the salience of the various traits, and x, y, and z are parameters that provide for differences in focus on the different components (Tversky 1977; Kintsch 2014; Parker 2015; Blumson 2017). Such parameters have been usually defined as "cultural traits" or minimal units of cultural transmission (Naroll 1964; Pocklington and Best 1997; Borgerhoff et al. 2006; O'Brien and Lyman 2003; O'Brien et al. 2013). Virtually any clumping of culture has been regarded as a trait, from whole subsistence efforts to decorative elements on a moccasin. The usefulness of the concept is that it functions as a placeholder in the analyst's thinking, signifying the "lowest level of cultural content" that the analyst cares to consider at a given time for a given purpose (Gatewood 2001). They seem to be polythetic in the sense that they should be defined as n-dimensionally variable, permitting a variation approaching continuous gradation of similarity and difference in their distributions (Needham 1975; Gatewood 2001). In Axelrod (1997) words, not any observable trait constitutes "culture" but only *the set of individual attributes that are subject to social influence*. Eerkens and Lipo (2005) have added the condition that cultural traits represent any measurable unit that we can delineate within the observed social variation and that can be argued to have inheritance continuity.

Among certain scholars, there is a tendency to call *memes* to such minimal units of ideas, symbols, values, or practices transmitted from agent to agent on occasion of an interaction act (Dawkins 1976; Distin 2005). Richard Dawkins proposed such a term in *The Selfish Gene* (1976) as an instrumental concept in evolution studies to explain the diffusion of ideas and social phenomena. Examples of "memes" would be melodies, dittos, beliefs (specially, religious beliefs), fashion, technological knowledge, decorations, etc. Dawkins suggested "meme" as an abbreviation of Greek word μιμεμα ("something imitated"). In any case, he asserted that what he was looking for was a simple word sounding similar to "gen." "Memes" appear to have no size: can be constituted by a single word or a complete text. John S. Wilkins

(1998) retained the idea of "meme" as a kernel for cultural imitation, describing it as a bit of information to be copied from agent to agent, without considering whether such bit contained other information or was integrated in a higher "meme." Gurnek Bains has developed this idea introducing the concept of *cultural DNA*, defined as a collection of genetic instructions used in the growth, development, functioning, and reproduction of any culture. Such instructions are the unwritten rules called norms and the common set of values that over time defines how social action is done in a particular group of people (Bains 2015). In spite of the use of the biologically laden term, DNA, the focus is on the deeply grained aspects of a culture that are replicated over generations rather than biological differences (Lee 2016, 2018).

As analysts, we can build a metric cultural space to investigate the pattern of similarity in beliefs, values, behaviors, and symbols within a group of people and between differentiated groups. In mathematics, a metric space is a set for which distances between all members of the set are defined. Those distances, taken together, are called a metric on the set. Those spaces are defined in terms of their dimensions, informally defined as the minimum number of coordinates needed to specify any point within it. This mathematical construct has been used to define *conceptual spaces* as analogues to a Euclidean geometric structure that represents a number of quality dimensions denoting basic properties by which concepts and objects can be compared. In such a conceptual space, *points* denote objects, *regions* denote concepts, and *natural* categories are convex regions in conceptual spaces. Within such a metric structure, if *a* and *b* are elements of a region, and if *c* is between *a* and *b*, then *c* is also likely to belong to the region. The notion of concept convexity allows the interpretation of the focal points of regions as category prototypes, and concepts are defined in terms of similarity to their prototypes (Gärdenfors 2004; Augello et al. 2013; Kovács and Hannan 2015). Consequently, concepts can be defined "polythetically" by reference to a set of properties which are both necessary and sufficient (by stipulation) for membership in the specific region of the conceptual space (Needham 1975; Barsalou 1985; Medin 1989; Ellen 2003; Medin and Rips 2005). These regions in the conceptual space are "blurred" categories, because they are not fixed entities; rather, they are constructed on each usage by combining attribute values that are appropriate to the context.

Ethnic identity can also be studied in terms of a metric conceptual space; in that case, points are social agents, separated by "cultural distance" and defined by a finite number of "cultural" dimensions representing cultural norms, world views, attitudes, perceptions, and ideas (Raza et al. 2001; Shenkar 2001; Sousa and Bradley 2008; Shenkar 2012). Therefore:

1. Each social agent is characterized by a large (but unspecified) number of memes or cultural traits, represented as components of a vector G.
2. Each cultural trait is possessed by large numbers of social agent.
3. No cultural trait is possessed by every social agent, identified (by itself or by others) as a member of a distinct group.

Consequently, "cultures" or "ethnic groups" can also be defined "polythetically" and represented as convex regions in a metric cultural space defined by reference to a set of cultural dimensions (social beliefs, behaviors, and values) (Clarke 1968). These regions in the cultural space have no fixed or well-defined borders: we can arrange its members along a line in such a way that each individual resembles his/her nearest neighbors very closely and his/her furthest neighbors less closely. The members near the extremes would resemble each other hardly at all, e.g., they might have none of the components of G in common. The only criterion for defining the identity group is the universal clustering criterion: the most "different" individual within the group is more similar to any other member of this group than to any other individual out of the group.

Because it is a "polythetical" entity, no single property is necessary for membership in an identity group; and nothing warrants or rule out the possibility that some cultural trait be sufficient for defining identity. This definition contrasts with the traditional definition of cultures and ethnic groups by reference to a cultural trait which is necessary and sufficient for membership in its extension. The difference between "cultures" and "ethnic groups" would be stressed in temporal terms: "cultures" are temporary patterns of similarity and regularity, whose duration is estimated in less than a generation, whereas ethnic groups are long-term patterns of similarity and regularity, built by learning and transmission from one generation to the next.

Axelrod (1997) applied those ideas by building a $L \times L$ square lattice of cells, where each cell represents a stationary individual who is endowed with a certain culture. An individual's cultural space is characterized by a list of d dimensions. The length of the vector D represents the social complexity of the population (i.e., the larger D is, the greater the number of different social criteria an individual needs to assign him/herself to his/her group). For each dimension, there is a set of q traits, which are the alternative values the dimension may have. The larger q is, the larger the number of possible traits that a given feature can have, corresponding to a higher complexity in the identity. It is assumed that all agents share the same value for d and all dimensions have the same value q. Thus, according to Axelrod, individual i's culture is represented by a vector x_i of d variables, where each variable takes an integer value in the range $[0, q - 1]$. This model allows exploring the spatial distribution of emergent cultural regions: sets of spatially contiguous agents who share an identical vector of culture. Naturally, parameters d and q influence the probability with which the system evolves to a monoculture (only one cultural region) or to global polarization (several multicultural regions). Continuous interaction between culturally similar agents would result in a decrease in cultural diversity until fill homogenization. Nevertheless, the results are extremely sensitive to the number of features the agents hold. Other factors that may influence results are their complexity in terms of the number of features, and the size of the neighborhood (see also Gracia-Lázaro et al. 2009; Guerra et al. 2010; Pfau et al. 2013; Kandler and Sherman 2013).

Starting from an initial condition in which the cultures of the human groups existing in a territory are randomly generated and therefore tend to be ethnically different, there is in fact a process of progressive cultural homogenization, with neighboring groups converging on a common culture, but this process never reaches completion. Parameter q, which defines the possible attributes in each cultural dimension, can be seen as a measure of the initial disorder or cultural variety in the system. In Axelrod's model cultural assimilation takes place because in each cycle each agent randomly selects one of its neighbors and adopts one of the properties of this neighbor's culture with a probability which depends on the degree of similarity already existing between the two cultures. Therefore, if two neighboring groups have completely different cultures and have no tendency to change their cultures because they are already part of culturally homogeneous regions, the two groups will forever keep their different cultures, and there will never be complete cultural homogenization. One could explain Axelrod's results as due to the fact that the simulation reaches a "frozen" state in which no further change is possible because the culture of every human group is either completely identical or completely different from the culture of neighboring groups. Given the assumptions of Axelrod's model, in these circumstances there can be no further social influence and cultural assimilation.

Measuring Culture and Ethnicity in the Archaeological Record

An archaeological definition of "culture" cannot be reduced to documenting the presence of some particular cultural traits or memes in the archaeological record, but it should be defined in terms of the mechanism marking how a population has arrived to a particular level of similarity in a majority of its activities. What members of the same group share is not what matters but the fact that they have learnt the same than some other people, and as a consequence of this learning and sharing process through time, a similarity degree in belief and behavior has emerged.

We should measure in the archaeological record patterns of similarity that are to a certain degree very general and can be related with possible communalities in needs, motivations, goals, actions, behaviors, and mediating artifacts. The goal of an archaeological investigation of ethnical identity in the past is then to measure the degree of "regularity" in the material elements used to produce and to reproduce the social group. We will look then for similarities in size, shape, texture, materiality, and placement of produced goods of any kind – from "hut" to "hats" – and also the similarities in size, shape, texture, materiality, and placement of working instruments and activity areas. More than a mere "similarity" between archaeological artifacts, we need a similarity between behaviors and, if it were possible, "values" and social norms. Because a particular behavior cannot be individualized only from its observed material consequence, we need to rebuild the behavior using all information available: the product, the residues produced by the activity, wear traces, the context, the technology, the physics behind activity and technology, etc.

The first problem lies in the scale of analysis. We cannot evaluate individual behavior in archaeology because in most cases the minimum level of social organization observable at the archaeological level is the "household" or the "local group" (see discussion in Hayden and Cannon 1982; Dornan 2002; Lock and Molyneaux 2006; Barceló and Maximiano 2012; Carballo et al. 2014; Fogle et al. 2015; Mills 2017). That means that many aspects of individual identity are out of the analysis. The only solution is to use "places" as surrogates of social agents having acted there. We are looking for "households" or spatiotemporal aggregates of social activities that go well beyond the production of subsistence and integrate the production level, with maintenance and residential activity and social reproduction behaviors. The "household" is also a polythetical entity, with clear boundaries, and its definition will vary according to context.

There is a functional hierarchy of spatial units that can be used for an analysis of cultural regularity, from the hearth to the territory. The best approach is a multiscale analysis comparing similarity patterns among different spatial categories: after a comparison of hearths from different built structures and from different settlements, we proceed with a comparison of spatially distinct activity areas, built spaces of diverse kinds, settlements, etc. The purpose is to build a complex multidimensional relational structure where similarity relationships at one scale – defined over a particular list of features – are statistically compared to similarity relationships at higher scales, each one defined over its particular list of features.

To calculate the similarity between those spatial units, we can start with an n-by-m places-by-behavior evidence matrix X, in which cell x_{ij} gives the evidence found at i about the occurrence of behavior j. We need a separate matrix for each spatial scale, in such a way that hearths should be compared with hearths, huts with huts, households with households, taking into account the problems of identification when comparing buildings with different functions. In the same way, the list of archaeological observations evidencing the occurrence of a particular behavior at that place should be the more exhaustive possible and well characterized in terms of place, product, technology, and indirect consequences. Matrix entries are then categorical choices expressing the presence of particular attribute for each of the variables. It is then an equivalent of Axelrod's way of measuring culture as a list of d dimensions representing complexity of the population (i.e., the larger D is, the greater the number of variables necessary to functionally analyze a particular place). For each dimension, there is a set of q traits, which are the alternative values the dimension may have. The larger q is, the larger the number of possible traits that a given feature can have, corresponding to a higher complexity in the identity. It is assumed that all agents share the same value for d and all dimensions have the same value q. The analysis begins by constructing a place-by-place agreement matrix M in which m_{ij} equals the number of behaviors from the list that are attested simultaneously at i and j. There are many alternative ways to build such a similarity matrix (Santini and Jain 1999; Cha 2007; Janowicz et al. 2008; Pirró 2009; Choi et al. 2010; Barceló 2015; Abbaspour et al. 2017). In case of simple presence/absence variables, we can use standard Hamming distances to compare overlap between both spatial units (Norouzi et al. 2012). The Hamming distance between two strings of equal length is the number of positions at which the corresponding elements are different.

$$d^{\text{HAD}}(i,j) = \sum_{k=0}^{n-1}\left[y_{i,k} \neq y_{j,k}\right]$$

In the equation d^{HAD} is the Hamming distance between the agents i and j, k is the index of the respective variable reading y out of the total number of variables n. In this way, agents measure the minimum number of *substitutions* required to change one string into the other or the minimum number of *errors* that could have transformed one string into the other. In our case, if cultural traits are binary (1 = "equal," 0 = "different"), calculating the cultural distance between a and b, the Hamming distance is equal to the number of ones (the traits count) in a XOR b. For example:

A	0100101000
B	1101010100
A XOR B	1001111100

The Hamming distance (H) between these 10-bit strings is 6, because this is the number of "discordances" between both vectors (the number of 1's in the XOR string). This result can be normalized to the range [0, 1], which is known as "*p*-distance," by dividing the resulting number by 10 (length of the vector).

In case we want to take into account the different number of items signaling to the same activity, we can use Morisita-Horn similarity index (Morisita 1962; Horn 1966):

$$C_H = \frac{2\sum_{i=1}^{S} x_i y_i}{\left(\dfrac{\sum_{i=1}^{S} x_i^2}{x^2} + \dfrac{\sum_{i=1}^{S} y_i^2}{y^2}\right) XY}$$

where

- x_i = number of times the item i has been signaled in spatial unit X, as an evidence of social activity I.
- y_i = number of times the item i has been signaled in spatial unit Y, as an evidence of social activity I.
- S = number of unique items.

$C_H = 0$ if the two samples do not overlap in terms of the observed archaeological evidence of the same activity and $C_H = 1$ if the activities occurred in the same proportions in both units.

The resulting similarity or distance matrix is then subjected to a principal factor analysis or correspondence analysis. It results in a set of eigenvectors and associated eigenvalues, or set of *inertia values for each correspondence axis, which are* the sum of squares of the singular values, i.e., the sum of the *eigenvalues*. The eigenvalue or inertia values can be used to assess the extent to which agreements among social agents are explained by a single factor, corresponding to the existence of a single cultural domain. For instance, consider a number of 60 possible behaviors, enumerating all the possible ways of decorating a pot, with only 2 possible choices: present and absent ($L = 2$). A total of 200 individualized places functionally equivalents (built spaces with a hearth and surrounded by stone wall) are used to analyze cultural consensus at that theory, yielding a 200*60 response matrix X. After forming the chance-corrected agreement matrix, we run a correspondence analysis to obtain eigenvalues (or the inertia of each axe). In case the first axe or factor is times larger than the next largest, the pattern would be highly consistent with the assumption of internal regularity – a prerequisite for ethnic identity. We should compare those results with alternative functional places and alternative lists of documented behaviors. When the first eigenvector turns out to be sufficiently dominant, we can go ahead and interpret the factor loadings (the values of the eigenvector) as estimates of each social agent knowledge of what it shares with the rest of the group.

We do not have enough with similarity calculations to understand the emergence of ethnic identity. Out of a group of related social agents, if two share the majority of cultural traits or "memes," can we reasonably hypothesize that they belong to the same ethnic group? In many cases the answer will be no. Overall similarity may be misleading because there are actually two reasons why social agents have similar characteristics and only one of them is a consequence of common identity. Using the standard vocabulary from biology, we can refer to inherited traits that appear to be similar in two populations as homologous feature (or homology). On the other hand, when unrelated individuals adopt a similar way of life, their activities may end up resembling one another due to efficiency matters or mere occasional convergence. Again, using the vocabulary from biology, when two individuals have a similar characteristic because of convergence, the feature can be referred as an analogous feature (or homoplasy). Only homologous similarity is evidence that two social agents are ethnically related. However, if two social agents share the highest number of homologies, can we reasonably assume they belong to the same group? The answer is still no – a homology may be a consequence of functional efficiency; only *inherited* homologies (called in biology synapomorphies) are evidence that ancestors and descendants are closely related so that they share the same identity. Associations are likely to be learned if they involve properties that are important by virtue of their relevance to the goals of the system. Therefore, because ethnicity is an intrinsically historical mechanism, we need information about the historical process of divergence or resemblance. The biological metaphor we have just presented can also be approached using a physical analogy, suggesting the concept of *social inertia* or resilience between different temporal states of the same aggregate or population. Social inertia has been defined as the ability of an aggregate of social agents to maintain a certain identity in the face of historical change and external perturbation (Carpenter 2000; Ramasco 2007; Kandler and Sherman 2013). An advantage of archaeology over other forms of cultural analysis is the possibility of analyzing change through time.

It can be argued that only observed and measured shared derived characters between social agents or groups of social agents could possibly give us information about how ethnic identity has been built. Following again the biological metaphor, we can refer to the method that groups social agents that share derived characters as *cladistics*. The relationships can be shown in a branching hierarchical tree called a *cladogram*. The cladogram is constructed such that the number of changes from one character state to the next is minimized. The principle behind this is the rule of parsimony – any hypothesis that requires fewer assumptions is a more defensible hypothesis

The first step in basic cladistic social analysis would be to determine which cultural traits, "memes," normative behaviors, or symbols are primitive and which are derived. Let us distinguish between members of the group ("in-group") and agents that do not share the same identity, although may have something in common for other reasons ("out-group"). The only way a homologous feature could be present in both an in-group and an out-group would be for it to have been inherited by both from an ancestor older than the ancestor of just the in-group. Consider the following example in which a character has states "present" and "absent." There are only two possibilities:

1. The absence of that feature does not contribute to the cultural consensus shared by the group, while its presence is what determines the common culture.
2. The presence of that feature does not contribute to the cultural consensus shared by the group, while its absence is what determines the common culture.

If we observe the presence of that feature in an individual outside the group being studied, the first hypothesis will force us to make more assumptions than the second (it is less *parsimonious*). Therefore, hypothesis 2 is more parsimonious and is a more defensible hypothesis. This example illustrates why an out-group analysis gives the most parsimonious, and therefore logical, hypothesis of which state is not related with the identity core.

Cladistic analysis has been used extensively in archaeology (O'Brien and Lyman 2003; Brantingham and Perreault 2010; Tehrani 2011; Houkes 2012; O'Brien et al. 2013; Lycett 2015; Lipo 2017). It can be used for our purposes, but we need previous information about members of in-group and members of out-group. And this is not always easy to estimate using just archaeological observables: nothing assures that two neighboring settlements belong to the same "culture" or "ethnic group" on the basis only of its neighborhood!

Many studies of social inertia suggest the three basic assumptions for cladogenesis are not always relevant for understanding the formation of "cultural relatedness" between human populations:

1. Any group of social agents is *related by descent* from a previous group of social agents.
2. There is a *bifurcating pattern* of cladogenesis.
3. *Change* in characteristics occurs in lineages over time.

The first assumption essentially means that humans arose on earth only once, and therefore, all human beings are related in some way or other. This can be true from the biological point of view, but not culturally! Not all cultural vectors come from a common one. To maintain such assumption, we need to restrict the analysis to precise geographical areas and delimited periods of time, so that we analyze how a particular cultural vector *evolved* into a multiplicity of cultures.

The second assumption is even most controversial; an existing population in time 0 is not necessary divided into exactly two groups at successive time steps! Multiple new lineages can arise from a single originating population at the same time or near enough in time to be indistinguishable from such an event. While this assumption could conceivably have occurred somewhere, it is generalizable to any historical context.

The final assumption is the most important assumption in cladistics, and it seems pretty obvious in the cultural case. However, it implies the necessity of actually measuring the temporal position of each time step, its duration, and the temporal difference with successive time steps.

Building "Cultural" Consensus

We intend to create a computer simulation where agents representing social reproduction units (two adults and a number of descendants: a "household" or "family") can be "helped" to survive by "culturally similar" neighbours with enough amounts of labor and technology (Del Castillo et al. 2010, 2014). By doing so, the agent receives cooperation in the form of labor, raw material, or subsistence from selected agents in the neighborhood, what stablished a relationship of social influence. The higher the probability of survival due to labor cooperative activities, the higher the cultural similarity within the emergent group, and the higher the dissimilarity with agents out of the new cooperation network. Under this assumption, geographical distance weakens economic and social ties and can promote cultural differentiation. Code can be downloaded from the CoMSES Network-OpenABM (https://www.comses.net/codebases/f16c9d1c-8c90-42dd-9ef4-d2f5980ac8a8/releases/1.0.0/).

One time step (cycle or "tick") in the simulation roughly represents what happened in a region during one season; two cycles or ticks represent 1 year. Nine processes are responsible for all system dynamics: agents work for survival, and they use positive (exchange) or negative (robbery) interaction flows to compensate for circumstantial threats to survival. Consequently, they need to *identify* other agents and act accordingly. Identity evolves and updates, as a result of interaction.

Agents can be mobile hunter-gatherers or sedentary production units doing a farming economy that depends on local availability of seed, land, water, and sun. Additional factors affecting such economy are the amount of labor and availability of technology in the form of tools. Agents are involved in four kinds of activities: producing food, producing tools, exchanging food, exchanging raw material, exchanging tools, stealing food, and stealing tools. Produced food is expressed in

Table 1 Canonical PD payoff matrix

	Cooperate	Defect
Cooperate	R-c, R-c	S, T
Defect	T, S	Nothing, Nothing

energy terms that is to say in the same units as the survival threshold (kilocalories). Food is obtained by agent *i* by means of labor with the contribution of its own technology, used to compensate the local difficulty of producing food. Technology is expressed in an aggregated measure of efficiency, and not in terms of the number of tools. Tools should be manufactured, and the agent needs knowledge and raw material for that. In case the agent cannot produce enough food for subsistence, because the amount of land and labor or means of production is not enough, then the agent should look for alternative sources of food (EXCHANGE or ROBBERY). If after looking for those alternatives the produced food still remains below the threshold, the agent dies.

New tools can only be obtained through exchange or robbery. The number of new tools is an external parameter to the model. The number of regions which are rich on metal ores is another potential external parameter. Both can be selected for experimenting with different scenarios.

When an agent needs food or tools, it looks for someone from the same identity group to ask for help. When the agent with surplus receives a petition for help, it should decide wheter it sends surplus to someone in need or it should keep surplus for its own future consumption. It is important to take into account that sending food or any other good implies for the sender an important cost: it reduces its surplus and may affect to the probability of surviving at later steps. It can produce, however, a benefit in the long run, because it will increase cultural consensus and decrease the risk of conflict later in the future. To be rational, the decision whether sending the requested food or refusing the proposed exchange should imply a way to evaluate the advantages or drawbacks of this behavior. This can be implemented in terms of *utility threshold* to be maximized. *Utility* is "usefulness," the ability of something to satisfy needs or wants. More than mere "satisfaction," it represents in our case the expected benefits in the long term of interacting with another. The state space for this decision can be visualized in a 2 × 2 matrix which records each of the four possible outcomes as a duple (an ordered set of two elements) (Table 1).

Suppose that the two agents are represented by the colors red and blue and that each agent chooses to either "Cooperate" or "Defect." If both agents cooperate, they both receive the *reward*, *R*, for cooperating but assume its cost. If blue defects while red cooperates, then blue receives a *temptation*, *T*, *that in any case will be higher than R − c, the original reward minus its costs*. In this scenario, red always receives a negative payoff, *S*, because it should assume the cost of production alone. Similarly, if blue cooperates while red defects, then blue receives the negative payoff, *S*, while red receives the temptation payoff, *T*. If both players defect, they do not gain

Table 2 Agents in surplus and needs v.1

		Agent with surplus	
		Exchange	Refuse exchanging
Agent in need	Exchange	2, 2	0, 5
	Refuse exchanging	0, 2	0, 5

Table 3 Agents in surplus and needs v.2

		Agent with surplus	
		Exchange	Refuse exchanging
Agent in need	Exchange	5, 5	0, 0
	Refuse exchanging	0, 0	0, 0

nothing. In a classical prisoner's dilemma game, the following condition must hold for the payoffs:

$$T > R - c > Nothing > S$$

The payoff relationship $R - c > Nothing$ implies that mutual cooperation is superior to mutual defection, while the payoff relationships $T > R - c$ and $Nothing > S$ imply that defection is the dominant strategy for both agents. That is, mutual defection is the only strong Nash equilibrium in the game (i.e., the only outcome from which each player could only do worse by unilaterally changing strategy). The dilemma then is that mutual cooperation yields a better outcome than mutual defection, but it is not the rational outcome because the choice to cooperate, at the individual level, is not rational from a self-interested point of view.

Our case is a bit different, because when the agent is in need of subsistence, it is obliged to look for cooperation. It is the agent with surplus who should take the decision. It knows that if it accepts the proposed exchange act, although costly, because it loses some of its surplus, it will be rewarding because the agent that asks for help will help when asked later (Table 2).

A strict application of the logic of the prisoner's dilemma would suggest a preference for the agent with surplus to refuse exchanging, because there is no advantage in the present. This seems to be the best strategy at short term. It would allow retaining the amount of surplus the agent has obtained so far. In any case, the greater the ease with which an agent obtains needed resources, the more predisposed to help at no cost. This is because the more cooperation today, the more expectations to cooperate in a more or less near future.

However, the agent with a surplus in food should consider that if it refuses the exchange now, when it asks for need sometimes in the future, the region that actually needs to increase its food will take revenge: cultural consensus will have decreased, and the chances of being helped will be very low. However, the reward should be expected in the future, because by assuming the cost to help, it also reduces cultural distance and contributes to form a more solid identity group where internal conflict is minimized and the possibilities of mutual defense increased (Table 3).

Each entry in the new payoff matrix is best read as an "If ... and ..., then" statement. In this case, if the agent in need asks for food to allow its survival and the agents with surplus choose to increase their labor effort and send some of its surplus, even though it implies a cost, then both agents receive the same reward in the future: they increase cultural consensus what increases the probability of future cooperation. If the agent with surplus chooses to refuse exchange, then agent in need will receive 0 points in the future, but it will also receive no future benefit.

Payoffs may vary according to the perception of communality between agents. Obviously, the agent in need asks everyone, but the agent with surplus can use the perceived "culture" of the other agent to make more precise expectations of the commitment of the agent in a future interaction. The lesser the cultural distance, the lesser the risks of actual defection. The reward in the future will be higher in case both agents belong to the same "culture." Given that each agent has an IDENTITY built in the form of a cultural vector (structured set of "memes"), the feeling of belonging to the same group is estimated in terms of the Hamming distance between the agent with surplus current cultural vector and the current cultural vector of the agent in need. To decide whether both belong to the same group, a threshold is needed. In so doing we are following Valori et al. (2012) and Stivala et al. (2014) suggesting a *bounded confidence* variant of the Axelrod model, in which a threshold θ is defined, such that agents can only interact when their cultural similarity is greater than or equal to θ. If $\theta = 0$, then this is equivalent to the model without bounded confidence. The rationale is that agents need a minimum level of "common ground" to interact at all. If it is higher, then cooperation occurs; if it is lower, then there is a probability of negative cooperation: violence and robbery. *ST* represents the cultural distance that should be overcome to be able to cooperate. Only when $d^{HAD}(i, j) >$ ST, agents i and j share group. Conflict (robbery, violence) will appear with a probability that is always directly proportional to the difference between $d^{HAD}(i, j)$; therefore, the more probable I and j are within the same identity group, the less risk of being attacked.

This parameter is the exactly opposite to Schelling's tolerance (1978): the higher the number of needed common features to build cultural consensus s, the lesser the tolerance with "the other" difference. The easier to build cultural likelihood, that is, the less common traits are needed, the higher the tolerance (Gracia-Lázaro et al. 2011; Zhang 2015).

Once the agent exchanges for increasing the chance of its own survival, the identity vector is updated toward the statistical mode of the exchanging agents' identity. With a fixed probability level (95%), each agent copies the statistical mode of identities within the group. In this way, agents try to fit their respective individual identity to collective identity if economically advantageous. Adaptive, different forms of cultural consensus emerge by combining the identities and values of interacting agents in an emergent group.

Our simulation implements then a somewhat modified iterated prisoner's dilemma. Agents seriously consider whether to maintain the opportunistic decision of getting the highest payoff in the short term or to opt for what seem the best for all agents in the territorial network in the long run: cooperation and high enough cultural consensus. This is because the players know that there is a distinct chance that they will meet again. The likelihood of future interactions thus casts a shadow on the present and arouses the possibility of an altruistic strategy.

Validating the Model with Archaeological Data

To test whether our model of identity formation is historically correct, we need to compare simulated results with empirical measures of communality and social inertia. We use the same statistical indexes to explore simulation output and the archaeological empirical data. Here, what we intend to test is whether cultural differences between different temporal steps have appeared just by chance (Madsen and Lipo 2014; Crema et al. 2016; Madsen and Lipo 2016) or as a consequence of restricted cooperation in parochialistic communities (Centola et al. 2007; Townley et al. 2011; Stivala et al. 2016). A source of neutral identity change through time has been implemented in the simulation in the form of an INTERNAL CHANGE RATE (IRC) or random parameter that introduces a random mutation rate. This is a random value (from 0 to 1, usually very small) defined in analogy to the probabilities of internal change (invention, mutation, catastrophe, sudden change). Then every tick, and with a fixed probability level determined as an external parameter, the identity vector mutates.

Simulation output is just a matrix *SOCIAL AGENTS x ATTRIBUTE* of surviving agents at the end of each simulation cycle, with all their attributes: current cultural vector, sum of acquired energy, current value of technological efficiency, etc. We can define spatial clusters of agents having cooperated during that cycle and temporal clusters of agents having cooperated without interruption in the last cycles. In the same way, the archaeological data to be used for validation will have the appearance of a series of *SPATIAL UNITS x ATTRIBUTES* matrices, one for each kind of spatial unit (from individual hearth until the settlement), with all their attributes: archaeological consequences of the occurrence of a particular activity. The purpose of the analysis is to discover whether clustering in the simulation is analogous to clustering at the archaeological record.

A Global Measure of Cultural Proximity and Social Inertia. What is the expected average level of dissimilarity between two individuals drawn at random from the population? Bossert et al. (2011) and Kolo (2012) have introduced a generalized index of similarity, the generalized ethnolinguistic fractionalization index. Based on the identity vectors (the "culture") of each agent, a mutual similarity matrix between individuals takes the distance between them into account, comparing each individual within a society with each other individual and assigning them a similarity rank that ranges from zero (the two individuals are not similar across any dimension) to unity (the individuals are exactly the same on every dimension). In our case, this is the matrix of normalized Hamming distances or S_{rk} index between all agents at each time step, bounded between 0 and 1. Based on this matrix, the corresponding generalized resemblance value for a population with N individuals is given through:

$$G(S_N) = 1 - \frac{1}{N^2}\sum_{i=1}^{N}\sum_{l=1}^{N}S_{ij}$$

Obviously, every person is identical to themselves, producing the diagonal series of ones in the matrix (person 1 compared with person 1, person 2 compared with person 2, etc.). In addition, Bossert et al. stipulated reasonably that similarities must be transitive – person 1's similarity to person 2 must be the same as person 2's similarity to person 1. Hence, the matrix is symmetrical along the diagonal. In this example matrix, person 1 and person 2 have a similarity score of half, whereas person 3 is completely different from both person 1 and person 2. For instance, suppose a group of three members. The similarity matrix would be:

$$\begin{matrix} 1 & 0.5 & 0.25 \\ 0.5 & 1 & 0 \\ 0.25 & 0 & 1 \end{matrix}$$

The corresponding value of $G(S_N)$ is:

$$G(S_N) = 1 - (1 + 0.5 + 0.25 + 0.5 + 1 + 0 + 0.25 + 0 + 1)/9 = -0.38$$

$G(S_N)$ should be calculated for all the population but also within each emergent group, $G(S_i)$. Internal similarity ($G(S_i)$) can only be explained in terms of global similarity ($G(S_i)$).

We need such a measure both for the agents generated by the simulation, but also for agents having lived in the past, and whose archaeological remains are available. The problem with this calculation lies in determining the archaeological observability of different social agents. Whereas in the simulation we can observe and measure the activity of individuals, in the archaeological record, we only have access to packages of spatially aggregated behaviors that may be the result of the activity of a reduced group of agents. But we do not have evidence of the internal division between them. For instance, if we can detect an individual household in the archaeological record, we cannot distinguish the different activities of men, women, and children, nor their differences in identity. In the case of cemetery data, the situation is different, and we can calculate global measures of similarity in identity. Nevertheless, not all the population will be represented in the cemetery but a subsample with social rights.

There is no single solution to this problem but an interactive experimentation with different levels of aggregation in the simulation. In some cases, we can work with a simulation built around individual agents, and similarity measures can be compared with the same measurements in different cemeteries across a territory. In other cases, we should build virtual families aggregating social agents and compare those results with an analysis at the household level. The same would be possible at lower resolutions.

The particular relationship between cooperating agents and how cooperation increases identity can be explored using standard tools of network analysis and graph theory. If cooperating agents are represented as connected nodes, we can measure the *size* for such a network – the number of edges. Its *density* is defined as a ratio of the number of edges E to the number of possible edges in a network with N nodes, given by $D = 3(E - N + 1)/N(N - 3) + 2$. When the flow from the agent with surplus to the agent in need is taken into account, the equation would be: $D = T - 2N + 2/N(N - 3) + 2$. An average path length (or characteristic path length) can be calculated by finding the shortest path between all pairs of nodes, adding them up and then dividing by the total number of pairs. The diameter of the network will be defined as the longest of all the calculated shortest paths in a network. It is the shortest distance between the two most distant nodes in the network. It is representative of the linear size of a network (Wasserman and Faust 1994; Borgatti et al. 2009; Scott 2017). Especially interesting is the study of the spread of identity degree in such a complex network. In our case, it is an example of conserved spread, because the total amount of content that enters a complex network remains constant as it passes through.

To measure social inertia, we need comparisons through time. The simulation starts with all agents having the same identity vector (start-up condition) that varies through the simulation because of the internal change rate, positive cooperation, and the consequences of conflict. We can compare the current state of the identity vector with the identity vector at start-up using the previously defined Hamming distance. Because each social agent is located at a particular place, and x and y coordinates are known, we can spatially interpolate such a present-past distance using kriging or any other alternative method (Fig. 1).

This statistic gives information about the process of spatial aggregation, and more information can be extracted by analyzing the pattern of geographical distances between points with the same degree of similarity with the original vector.

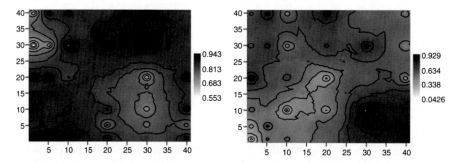

Fig. 1 Temporal evolution of social inertia through time. Interpolated values in gray scale where white represents zero similarity with start-up and black represents full similarity between identity at the beginning and at current time

It is important to take into account that locations in the map with the same degree of social inertia – similarity with the past – do not have necessary the same identity in the present.

This calculation is simple and very informative in the case of the simulation. It can be more difficult in the archaeological record. Because it is a comparison between agents at current time and a specific starting situation, we need to determine in the archaeological record an initial situation from which the current record historically derives. This is only possible if we have correctly asserted the temporality of different periods, and we know in detail the historical process of change and evolution.

Additionally, we can build a cladogram of all living agents at the current cycle, expressing the historical dependence between any pair of social agents. Such graph uses lines that branch off in different directions ending at a group of agents with a last common ancestor. In biology it is a *clade*; in our case, it corresponds to an "ethnic group," according to the definition presented at the beginning of the paper. There are many shapes of cladograms, but they all have lines that branch off from other lines. The lines can be traced back to where they branch off. These branching off points represent a hypothetical ancestor (not an actual entity) which can be inferred to exhibit the traits shared among the terminal groups above it. This hypothetical ancestor might then provide clues about the order of historical change of various features. In such a cladogram, we can measure paths and make calculations that will give some cues about social inertia. We can use a measure of cultural proximity that has already been used in cross-cultural and linguistic research (Eff 2004):

$$S_{rk} = \frac{\partial_x - \partial_{rk} + 1}{\partial_x + 1}$$

where S_{rk} is the similarity between the identity vector ("culture") of any two social agents, ∂_x is the length of the longest path in the emergent group (i.e., the length of the longest path to the common ancestor of the entire population, that is the common identity vector shared for all agents at start-up), and ∂_{rk} is the length of the longest path to the nearest common ancestor of agent r and agent k. Thus, if the path separating two agents in the present cycle has a length of 5, and the path to the start-up vector is 10, then $(10 - 5 + 1)/(10 + 1) = 0.5455$. If there is no other near ancestor except the vector at start-up, then $(10 - 10 + 1)/(10 + 1) = 0.0909$. Self-similarity is $(10 - 0 + 1)/(10 + 1) = 1.0$. Proceeding in this way, one can calculate a proximity measure between each pair of agents within the same ethnic group. In case there are no links between rival groups, agents will have a similarity of zero with agents outside its own ethnos. The similarity between each member of a group will thus always lie between zero and one. Self-similarity= 0 only when the two agents belong to different groups and will equal to 1 only when a language is compared with itself.

Social Fractionalization This is a measure of the degree the population is divided into groups with different identity. It implies dividing the population into ethnic groups, calculating each group's share of the population, summing the squared shares, and subtracting the sum from one.

To calculate the number of groups, we need a standard clustering of identity vectors according their joint similarity or distance. We can use the Hamming distance or the previously presented S_{rk} index between identity vectors to build a similarity matrix between agents. The optimal number of groups can be calculated as follows (Thorndike 1953):

1. Compute clustering algorithm (e.g., k means clustering) for different values of k. For instance, by varying k from one to ten clusters.
2. For each k, calculate the total within-cluster sum of square (*wss*).
3. Plot the curve of *wss* according to the number of clusters k.
4. The location of a bend in the plot is generally considered as an indicator of the appropriate number of clusters.

To determine the quality of this clustering, it is necessary to determine how well each agent belongs to its group. Suzuki and Shimodaira (2015) suggest using bootstrap resembling techniques to *compute a p-value* for each *hierarchical cluster*. The method implies the random generation of thousands of bootstrap samples by randomly sampling elements of the data; then a new hierarchical clustering is computed on each bootstrap copy. For each cluster we should compute (1) the *bootstrap probability* (*BP*) value which corresponds to the frequency that the cluster is identified in bootstrap copies and (2) the *approximately unbiased* (AU) probability values (p-values) by multiscale bootstrap resampling. Clusters with AU >= 95% are considered to be strongly supported by data.

Given that we know the spatial coordinates x and y of each agent, by performing a nonmetric multidimensional scaling of respective identity vectors, we can reduce the dimensionality of the underlying conceptual space (ten dimensions) to just one and map the scores of each agent on each coordinate using the original x and y location. The result is a map of similarities, showing where in the territory concentrates agents with the same identity.

Once calculated the number of groups, and given mutual exclusiveness and exhaustiveness, we should measure the probability that two randomly chosen individuals from a neighborhood's population belong to different groups. The measure should score zero in a perfectly homogenous population (i.e., all individuals in the population belong to the same group) and should reach its theoretical maximum value of 1 where an infinite population is divided into infinite groups of one member (Alesina et al. 2003).

Such a measure has been calculated by Taylor and Hudson (1972) as a decreasing transformation of the Herfindahl concentration index applied to population shares. In particular, the index takes the form of:

$$E_{(p)} = 1 - \sum_{k=1}^{k} p_k^2$$

where p_k^2 is calculated as the square of the number of agents belonging to the same group k divided by the total number of agents in the population at each time step. This is exactly the Blau's index of heterogeneity, also referred as Gibbs-Martin index in sociology, psychology, and management studies (Blau 1977; Sampson 1984; Blau and Schwartz 1997; Castellano et al. 2009). The same index has been proposed by the linguist Greenberg (1956) who termed it the *A index*. In the statistical literature, it is known as the Gini-Simpson index, introduced first by Gini in 1912 and then by Simpson in 1949 as a measure of diversity of the multinomial distribution.

Using the above equation, when there is no emerging groups and all agents behave autonomously, $E(p) = 1 - 0 = 1$. That means that if we consider any pair of agents at random, the probability they belong to different ethnic groups is 1, because each agent constitutes its own ethnic group. When all agents belong to the same group – the situation at the beginning of the simulation – $E(p) = 1 - 1 = 0$. That means that if we consider any pair of agents at random, the probability they belong to different ethnic groups is 0, because all the population has the same identity. This formula requires the groups to be mutually exclusive (i.e., if an agent is in group g, then it is not in group h) and exhaustive. Isolated agents only intervene in the calculation of the total population number.

This index of fractionalization is just a measure of heterogeneity; such measure conveys no information about the *depth* of the divisions that separate members of one group from another, which is a necessary factor for inferring *social tension* from mere fractionalization (Fearon and Laitin 2003; Posner 2004; Chandra and Wilkinson 2008; Brown and Langer 2010; Chakravarty and Maharaj 2011). Obviously, if dissimilarity is great and fractionalization is intense, the probability of competition should be higher. But the number of groups and the degree of difference on their own are not enough to conclude social tension and violence. The idea of "polarization" is needed to transform difference into competition. Theoretically, polarization should be calculated in terms of the "distance" between two groups, i and j, corrected by the sizes of each group in proportion to the total population (Esteban and Ray 1994; Duclos et al. 2003). The assumption behind this alternative measure is that while the generalized fractionalization matrix rightly attributes a low chance of ethnic conflict to a homogeneous population, highly fractionalized populations are not conflictual as no group has the "critical mass" necessary for conflict. Conflict will be more likely the more a population is polarized into two large groups, well beyond a specific critical mass. Montalvo and Reynal-Querol (2002, 2005; Chakravarty and Maharaj 2011) have developed an index of demographic *polarization*:

$$\text{RQ} = 4\sum_{i=1}^{k}\sum_{j}^{k} p_i^2 p_j = 4\sum_{i=1}^{k} p_i^2 (1-p_i) = 1\sum_{i=1}^{k} \left(\frac{0.5-p_i}{0.5}\right)^2 p_i$$

p_i in the equation is the proportion of all agents currently alive who belongs to each superagent i. RQ employs a weighted sum of population shares. The weights employed in RQ capture the deviation of each group from the maximum polarization share 1/2 as a proportion of 1/2. Analogously to the index of fractionalization, underlying the formula for RQ is the implicit assumption that any two groups are either completely similar or completely dissimilar, and thus, the weights depend on population shares only. This index tends toward zero for very homogeneous and non-conflictive populations, i.e., with only one relevant group. However, with increasing group numbers, $E(p)$ and RQ show clearly different results. While $E(p)$ is an increasing function of the number of groups, RQ reaches its maximum with two equally sized groups (i.e., $i = 2, p_1 = 0.5, p_2 = 0.5$) and decreases afterward. It is the same to say that social heterogeneity and social conflict are not one and the same. Initially, one could think that the increase in diversity increases the likelihood of social conflicts. However, this does not have to be the case. In fact, many researchers agree that the increase in ethnic heterogeneity initially increases potential conflict but, after some point, more diversity implies inferior probabilities for potential conflict.

Preliminary Results

We have only preliminary results in the case of mobile hunter-gatherer virtual agents (Barceló et al. 2014, 2015; Barceló and Del Castillo 2015). Cooperation in a hunting-gathering band does not imply the transfer of subsistence, because what an agent acquires is limited to its current needs. Consequently, there is no surplus of food to be transferred, but there is always a surplus of labor not used when resources are rich enough and easily accessible with the current labor capability.

Preliminary results show that in a majority of hunting-gathering scenarios, cooperation in the form of shared labor does not increase the probability of survival (Fig. 2).

This result was clearly unexpected before we built our simulation. It is only partially validated using archaeological data, but we can check whether global resemblance – the $G(_{SN})$ index – is correlated with the amount of resources. That is, if we have a number of settlements of different chronologies, and some paleoecological data about the temporal variations in the average of a particularly useful resource, we can check whether global resemblance diminishes as soon as the resource average also diminishes in that geographical area.

Cooperation drastically depends on the distance over which social interaction can be defined. The amount of cooperation is inversely proportional to the distance between agents. The impact of interaction radius, which depends on transportation technology, is also of relevance. In case groups are able to move around small areas, the chance of finding a group culturally similar enough for cooperating is far less than if social agents travel long distances. Furthermore, mobility increases stochasticity in all simulated scenarios. That is, at each run of the same scenario (with the same values at the same parameters at start-up), the evolution of the population differs. This is a consequence of the increasing irregularity in agents'

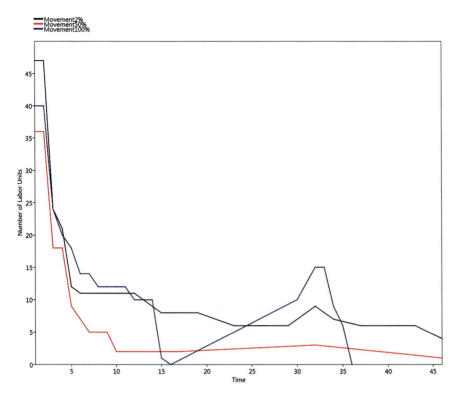

Fig. 2 Decreasing population of mobile cooperative individuals in poor world scenarios. Only one scenario with a mean of 6500 kcal at the warm season and 3250 kcal at the cold one. Resource irregularity fixed for an standard deviation = 1000 kcal

revenues. The mean energy acquired by labor unit is fairly constant in all simulated scenarios, but when adding mobility, its standard deviation also increases, varying enormously from one cycle to the other. That means that although most agents behave in the same way trying to extract the maximum amount of energy they could find locally, the local availability varies. We have fixed such an irregularity assuming a Gaussian distribution with a standard distribution of 1000 kcal. This value should be interpreted as a very small irregularity in the richest world (12.5% of variation) and increasing irregularity as the mean of resources is lower, arriving to 40% of variation in the poorest scenario.

If mobility increases stochasticity, then it cannot be interpreted as an adaptive decision to increase the expectances of survival. To move or not to move is no "prisoner's dilemma," because the agent plays against "Nature" (or against itself in a possible alternative situation), and not against other agent.

On the other hand, in most simulated scenarios, technological efficiency experienced changes and evolution. Here computational results can be easily compared with archaeological data to look for whether there is evidence of small but continuous changes, interpreted as local advances not related with interaction but also a gradual convergence toward the most efficient, when innovations diffused (see Barceló et al. 2015).

Because of economic interaction, virtual households aggregate in space, configuring what we can consider social networks of cooperation. In some contexts, closed groups may emerge, but when the intensity of interaction varied, or the circumstances in which the interaction took place were different, the nature of the social aggregation was also different, allowing the dissolution of any previously differentiated group into an undefined consensus. Social aggregation and network formation may have been more frequent when resources were lower and the need for surplus labor is higher. When resources are enough for the actual labor force at the group, benefits of cooperation are less obvious, and therefore, the probability of any form of restricted territoriality is significantly lower.

Our simulation shows that very few groups will keep moving again and again. Rather, some kind of "good-enough" scenario is found where groups stay in the neighborhood of other groups, keeping the connections among them.

Are emerging cooperation networks between culturally similar social agents an initial form of ethnogenesis? We stressed at the beginning of this paper that *the lesser the intensity and frequency of inter-group relationships, the greater the differences in ways of speaking and other cultural features manifested by groups.* The same can be said in terms of network embeddedness. Network embeddedness means that everybody does not interact equally with everybody else but is constrained by needs (expected benefits), geographical neighborhood, and prior cultural consensus (common history). Agents within the network interact among themselves more often than with others out of the network, which means that a subset of the population may be excluded from positive interaction and hence the process of similar identity negotiation and innovation diffusion.

Our model estimates that the probability that two randomly drawn individuals from the population belong to two different groups increase when resources are low and survival may be at risk. The higher the value, the higher horizontal inequality in the total population. Our preliminary results clearly show that when the simulated world is comparatively poor (maximum resources less than 30,000 kilocalories for a complete season), fractionalization scores are higher than in the case where resources are abundant and frequent. A high value of fragmentation when resources were scarce and concentrated can be explained as the probability of an individual's willingness to spend on available resources given the degree of affinity within its constrained neighborhood. $G(S_N)$ scores never remain constant. The probabilities of successful economic interaction vary depending on how many members of the community share the same identity of that individual.

When human groups reduce their residential mobility, settlements may be confined to locations with enough availability of critical resources (water, wood) and good condition (repair, mild winters). Parallel to this reduction in residential mobility, the ranges for logistic action would have expanded and extended, but there is also intuitively a much greater risk of social tension and competition when social agents are concentrated in one particular region of the country than if it were dispersed evenly across the country. In the hunter-gatherer-explored scenarios, the distribution of social aggregates is far from a simple bipolar case. Demographic polarization attains higher values when the world has the more abundant resources and when fractionalization has low values because most agents belong to a few groups. These results are different then to the expected increased territoriality as a consequence of resource scarcity and spatial concentration. It seems as if when economic mobility is great – hunter-gatherers – social aggregation is very low, and the size of groups is too reduced to allow for the emergence of social tension, segregation, and hence exclusive territoriality. On the other hand, when social networks are big enough to integrate a big number of previously isolated agents, social tension emerges between network-embedded individuals and people without any ascription.

In our simulation, virtual groups never configured territories with clear-cut Euclidean boundaries and explicit segregation. Our results stress the role of "territoriality" in terms of network embeddedness (Kim 2009); a "fractal metaphor" helps us to recognize that social aggregates overlapped. There was no place for delimited spaces conceived in geopolitical terms, because households aggregated in groups which had no "natural" limits.

Conclusions

Archaeologists do not dig for cultures and/or ethnicities. We think that "culture" and "ethnicity" should be understood as the propensity or tendency in the probability that a known social goal, motivation, behavior, or artifact be put on practice. Culture relates groups of individuals doing the same things, in the same way and with the same instruments in the present, whereas ethnicity relates individuals persisting in their similar action through time: they share historical group trajectories. When conscience about their own similarity emerges, the contexts in which this resemblance is expressed and reproduced become socially and historically structured, while at the same time, it constrains future activity. It is easy to see that this process of constructing similarity and communality of action is in constant evolution and transformation.

By emphasizing the need to build measures of social regularity and/or resilience through space and time, we insist in the absence of discrete groupings with clearly defined borders and frontiers but a network of logical relationships. Such relationships can be described metrically, with different possible topologies: in some contexts, closed groups may emerge, but when mechanisms of reproduction and interaction are different, or the circumstances on which they act change, the result can be very different, like the dissolution of any previous differentiated group into an undefined homology of social activities. Therefore, we consider that there are not ethnic groups as discontinuous isolates to which people naturally or "ideally" belong but a series of real nesting dichotomizations of inclusiveness and exclusiveness (social fusion and fission) resulting from social reproduction, that affect the way people aggregated and aggregate into groups and adapted/adapt their social practice in consequence. Ethnicity does not presuppose the existence of discrete and particular "ethnia", nor does culture imply the existence of cultures.

Not any measure of cultural consensus or similarity helps to explain why consensus exists and characterize a particular social group; they simply facilitate the discovery and description of human aggregates, whose aggregation mechanism is their common identity. We intend to formulate a hypothetical mechanism that generates cultural consensus in some particular historical scenarios. Our hypothesis is based on the idea that social agents tend to interact with agents with a similar "identity," what makes for a greater probability interactions between already connected people than unconnected ones (with dissimilar features). In addition, we also introduce the principle of social influence or confluence (Tang et al. 2013).

The concept of ethnic identity used in this chapter is understood as a methodological concept that allows to understand how the patterns of diversity and difference can be explained archaeologically. We refuse the classical substantialist and culturalist view of ethnicity that defined closed social groups on the basis of apparently visible features such as language, political organization, territory, and kinship as prior classification of cultural units and their limits. This conception of ethnicities considers the empirical existence of a predictable and systematic correspondence between distinctive cultural features and ethnic identities. We consider instead an *instrumentalist* approach, which analyzes identity in terms of a historical process with the participation of the rational action of individuals, the intentionality of actions, the calculation of perceived benefits, and the maximization of benefits for political and material purposes.

We have considered the construction of communalities more important than the mere definition of a group based on its limits and possible borders. Recognizing ethnicity as a social outcome constitutes one step forward to the fact that social interactions are part of historical processes. Ethnicity is a consequence of social interaction at a level above the single individual. At the highest levels, it involves situations of contrast and/orconfrontation between groups of individuals. If ethnicity is a social output, it implies consensus and complementarity relationships, ascriptions, and autoadscriptions, but these dynamics not always account for the political or economic context in which they are developed, neither the context of production of the social system that involves them.

Acknowledgments The research presented in this paper has been founded by research grants HAR2016-76534-C2-1-R, ECO2017-83534-P, and CSD2010-00034/G6 from the Ministerio Español de Economía y Competitividad, by the Generalitat de Catalunya (2017SGR243), and the Institut Català de Recerca Avançada (ICREA). We thank our colleagues Ricardo del Olmo and David Poza, from Universidad de Burgos for their help in programming a preliminary version of our model.

Bibliography

Abbaspour, R. A., Shaeri, M., & Chehreghan, A. (2017). A method for similarity measurement in spatial trajectories. *Spatial Information Research, 25*(3), 491–500.

Abruzzi, W. S. (1982). Ecological theory and ethnic differentiation among human populations. *Current Anthropology, 23*, 13–35.

Alba, R. (1990). *Ethnic identity: The transformation of white America*. New Haven: Yale University Press.

Aldén, L., Hammarstedt, M., & Neuman, E. (2015). Ethnic segregation, tipping behavior, and native residential mobility. *International Migration Review, 49*(1), 36–69.

Alesina, A., Devleeschauwer, A., Easterly, W., Kurlat, S., & Wacziarg, R. (2003). Fractionalization. *Journal of Economic Growth, 8*(2), 155–194.

Augello, A., Gaglio, S., Oliveri, G., & Pilato, G. (2013). An algebra for the manipulation of conceptual spaces in cognitive agents. *Biologically Inspired Cognitive Architectures, 6*, 23–29.

Axelrod, R. (1997). The dissemination of culture: A model with local convergence and global polarization. *Journal of Conflict Resolution, 41*(2), 203–226.

Bains, G. (2015). *Cultural DNA: The psychology of globalization*. Hoboken: John Wiley & Sons.

Barceló, J. A. (2015). Measuring, counting and explaining: An introduction to mathematics in archaeology. In J. A. Barceló & I. Bogdanovic (Eds.), *Mathematics and Archaeology* (pp. 3–64). Boca Raton, FL: CRC Press.

Barceló, J. A., Del Castillo, F., Del Olmo, R., Mameli, L., Miguel Quesada, F. J., Poza, D., & Vilà, X. (2014). Social interaction in hunter-gatherer societies: Simulating the consequences of cooperation and social aggregation. *Social Science Computer Review, 32*(3), 417–436. https://doi.org/10.1177/0894439313511943.

Barceló, J. A., Del Castillo, F., Del Olmo, R., Mameli, L., Miguel Quesada, F. J., Poza, D., & Vila, X. (2015). Simulating patagonian territoriality in prehistory: Space, frontiers and networks among hunter-gatherers. In G. Wurzer, K. Kowarik, & H. Reschreiter (Eds.), *Agent-based modeling and simulation in archaeology* (pp. 243–289). Berlin-New York. Advances in geographic Information Science: Springer-Verlag. https://doi.org/10.1007/978-3-319-00008-4__10.

Barceló, J. A., & Del Castillo, M. F. (2015). *Simulating prehistoric and ancient worlds*. New York/Berlin: Springer Verlag.

Barceló, J. A., & Maximiano, A. (2012). The mathematics of domestic spaces. In M. Madella, G. Kovács, B. Berzsenyi, & I. Briz (Eds.), *The archaeology of household*. Oxford: Oxbow Books.

Barsalou, L. W. (1985). Ideals, central tendency, and frequency of instantiation as determinants of graded structure in categories. *Journal of Experimental Psychology: Learning, Memory, and Cognition, 11*(4), 629.

Barth, F. (1969). *Ethnic groups and boundaries: The social organization of culture difference*. London: George Allen & Unwin.

Bate, F. (1998). Sociedad concreta y periodización tridimensional. In *Boletín de antropología americana*. D.F. México: Instituto Panamericano de Geografía e Historia.

Bellón Ruiz, J. P., García, F. M., & Álvarez, F. J. (2009). Pueblos, culturas e identidades étnicas en la investigación protohistórica de Andalucía (I). En F. Wulff, M. Álvarez (Eds.), *Identidades, culturas y territorios en la Andalucía prerromana* (pp. 51–74). Málaga: Universidad de Málaga.

Bentley, G. (1987). Ethnicity and practice. *Comparative Studies in Society and History, 29*(1), 24–55.

Bernardi, L. (2003). Channels of social influence on reproduction. *Population Research and Policy Review, 22*(5–6), 527–555.

Blau, P. M. (1977). *Inequality and heterogeneity: A primitive theory of social structure* (Vol. 7). New York: Free Press.

Blau, P. M., & Schwartz, J. E. (1997). *Crosscutting social circles: Testing a macrostructural theory of intergroup relations*. Milton Park (Oxford):Routledge.

Blumson, B. (2017). Two conceptions of similarity. *The Philosophical Quarterly, 68*(270), 21–37.
Bogdanović, I. (2011). La instrumentalització del passat en el present. La construcció de les identitats col.lectives dels Balcans centrals en la història de l'arqueologia sèrbia. PhD Dissertation. Universitat Autònoma de Barcelona.
Bongaarts, J., & Watkins, S. C. (1996). Social interactions and contemporary fertility transitions. *Population and Development Review, 22*(3), 639–682.
Borgatti, S. P., & Halgin, D. S. (2011) Mapping culture: Freelists, pilesorting, triads and consensus analysis. walnut creek. *The Ethnographer's Toolkit*. Available at: http://works.bepress.com/daniel_halgin/5/. Downloaded on 16 April 2018.
Borgatti, S. P., Mehra, A., Brass, D. J., & Labianca, G. (2009). Network analysis in the social sciences. *Science, 323*(5916), 892–895.
Borgerhoff Mulder, M., Nunn, C. L., & Towner, M. C. (2006). Cultural macroevolution and the transmission of traits. *Evolutionary Anthropology, 15*, 52–64.
Bossert, W., D'Ambrosio, C., & La Ferrara, E. (2011). A generalized index of fractionalization. *Economica, 78*, 723–750.
Bowles, S., & Gintis, H. (2004). The evolution of strong reciprocity: Cooperation in heterogeneous populations. *Theoretical Population Biology, 65*, 17–28.
Boyd, R., & Richerson, P. (2005). *The origin and evolution of cultures*. Oxford: Oxford University Press.
Brantingham, P. J., & Perreault, C. (2010). Detecting the effects of selection and stochastic forces in archaeological assemblages. *Journal of Archaeological Science, 37*(12), 3211–3225.
Brather, S. (2004). Ethnische Interpretationen in der frühgeschichtlichen Archäologie. *Geschichte, Grundlagen und Alternativen, 78*.
Braun, D. P. (1991). Why decorate a pot? Midwestern household pottery, 200 B.C.–A.D. 600. *Journal of Anthropological Archaeology, 10*(4), 360–397.
Brown, G. K., & Langer, A. (2010). Horizontal inequalities and conflict: A critical review and research agenda. *Conflict, Security & Development, 10*(1), 27–55.
Carballo, D. M., Roscoe, P., & Feinman, G. M. (2014). Cooperation and collective action in the cultural evolution of complex societies. *Journal of Archaeological Method and Theory, 21*(1), 98–133.
Carley, K. (1991). A theory of group stability. *American Sociological Review, 56*(3), 331.
Carpenter, S. (2000). Effects of cultural tightness and collectivism on self-concept and causal attributions. *Cross-Cultural Research, 34*(1), 38–56.
Casella, E. C., & Fowler, C. (2005). *The archaeology of plural and changing identities: Beyond identification*. Springer Science: Boston.
Castellano, C., Fortunato, S., & Loreto, V. (2009). Statistical physics of social dynamics. *Reviews of Modern Physics, 81*(2), 591.
Centola, D., González-Avella, J. C., Eguíluz, V. M., & San Miguel, M. (2007). Homophily, cultural drift, and the co-evolution of cultural groups. *Journal of Conflict Resolution, 51*(6), 905–929.
Cha, S. H. (2007). Comprehensive survey on distance/similarity measures between probability density functions. *City, 1*(2), 1.
Chakravarty, S., & Maharaj, B. (2011). Measuring ethnic polarization, social choice and welfare, Springer. *The Society for Social Choice and Welfare, 37*(3), 431–452.
Chandra, K., & Wilkinson, S. (2008). Measuring the effect of "ethnicity". *Comparative Political Studies, 41*(4–5), 515–563.
Choi, S. S., Cha, S. H., & Tappert, C. C. (2010). A survey of binary similarity and distance measures. *Journal of Systemics, Cybernetics and Informatics, 8*(1), 43–48.
Clark, W. A., & Fossett, M. (2008). Understanding the social context of the Schelling segregation model. *Proceedings of the National Academy of Sciences, 105*(11), 4109–4114.
Clarke, D. L. (1968). *Analytical archaeology*. London: Methuen.
Cosgrove, D., & Jackson, P. (1987). New directions in cultural geography. *Area, 19*, 95–101.
Crema, E. R., Kandler, A., & Shennan, S. (2016). Revealing patterns of cultural transmission from frequency data: Equilibrium and non-equilibrium assumptions. *Scientific Reports, 6*, 39122. https://www.ncbi.nlm.nih.gov/pmc/articles/PMC5156924/.

D'Andrade, R. (1987). A folk model of the mind. In D. Holland & N. Quinn (Eds.), *Cultural models in language and thought* (pp. 112–148). Cambridge: Cambridge University Press.

Dawkins, R. (1976). *The selfish gene.* New York: Oxford University Press.

Del Castillo, F., Barceló, J. A., Mameli, L., & Moreno, E. (2010). Etnicidad en cazadores-recolectores patagónicos: Enfoques desde la simulación computacional. *Revista Atlántica-Mediterránea de Prehistoria y Arqueología Social (RAMPAS), 12*, 35–58. Cádiz: Universidad de Cádiz. ISSN: 1138-9435.

Del Castillo, M. F., Barceló, J. A., Mameli, L., Miguel, F., & Vila, X. (2014). Modeling mechanisms of cultural diversity and ethnicity in hunter- gatherers. *Journal of Archaeological Method and Theory, 21*, 364–384. J. Skibo y C. Cameron. (Eds). US: Springer.

Demoule, J. P. (1999). Ethnicity, culture and identity: French archaeologists and historians. Theory in French archaeology. *Antiquity, 73*(279), 190–198.

Deutsch, H. (1997). Identity and general similarity. *Philosophical Perspectives, 12*, 177–200.

Dietler, M. (1994). Our ancestors the gauls: Archaeology, ethnic nationalism, and the manipulation of celtic identity in modern Europe. *American Anthropologist, 96*(3), 584–605.

Distin, K. (2005). *The selfish meme* (p. 205). Cambridge: Cambridge University Press.

Dornan, J. L. (2002). Agency and archaeology: Past, present, and future directions. *Journal of Archaeological Method and Theory, 9*(4), 303–329.

Dow, M. M., Burton, M. L., Reitz, K., & White, D. R. (1984). Galton's problem as network autocorrelation. *American Ethnologist, 11*(4), 754–770.

Dressler, W. W., Borges, C. D., Balieriro, M. C., & Dos Santos, J. E. (2005). Measuring cultural consonance: Examples with special reference to measurement theory in anthropology. *Field Methods, 17*(4), 331–355. https://doi.org/10.1177/1525822X05279899.

Duclos, L., Vokurka, R. J., & Lummus, R. R. (2003). A conceptual model of supply chain flexibility. *Industrial Management & Data Systems, 103*(5), 446–456.

Eerkens, J., & Lipo, C. (2005). Cultural transmission, copying errors, and the generation of variation in material culture and the archaeological record. *Journal of Anthropological Archaeology, 24*(4), 316–334.

Eff, E. A. (2004). Does Mr. Galton still have a problem? Autocorrelation in the standard cross-cultural sample. *World Cultures, 15*(2), 153–170.

Efferson, C., Lalive, R., & Fehr, E. (2008). The coevolution of cultural groups and ingroup favoritism. *Science, 321*(5897), 1844–1849.

Ellen, R. F. (2003). The cognitive geometry of nature: A contextual approach. In *Nature and society*, Anthropological perspectives (pp. 113–134). New York:Routledge.

Emberling, G. (1997). Ethnicity in complex societies: Archaeological perspectives. *Journal of Archaeological Research, 5*(4), 295–344.

Esteban, J., & Ray, D. (1994). On the measurement of polarization. *Econometrica, Econometric Society, 62*(4), 819–851.

Fearon, J., & Laitin, D. (2003). Ethnicity, insurgency, and civil war. *American Political Science Review, 97*, 75–90.

Fernández Götz, M. A. (2008). *La construcción arqueológica de la etnicidad. Serie Keltia, 42*. Noia: Editorial Toxosoutos.

Fischer, M. (2007). Culture and cultural analysis as experimental systems. *Cultural Anthropology, 22*(1), 1–65.

Fogle, K. R., Nyman, J. A., & Beaudry, M. C. (2015). *Beyond the walls: New perspectives on the archaeology of historical households*. Gainesville: University Press of Florida.

García Fernández, F. J., & Bellón, J. P. (2009). Pueblos, culturas e identidades étnicas en la investigación protohistórica de Andalucía. In *Identidad, cultura y territorio en la Andalucía prerromana*. Málaga: Universidad de Málaga.

García Montalvo, J., & Reynal-Querol, M. (2002). Why ethnic fractionalization? Polarization, ethnic conflict and growth. UPF Economics and Business Working Paper No. 660.

Gärdenfors, P. (2004) *Conceptual spaces: The geometry of thought*. Cambridge (MA):MIT Press.

Garro, L. C. (2000). Remembering what one knows and the construction of the past: A comparison of cultural consensus theory and cultural schema theory. *Ethos, 28*(3), 275–319.

Gatewood, J. B. (2001). Reflections on the nature of cultural distributions and the units of culture problem. *Cross-Cultural Research, 35*(2), 227–241.

Geertz, C. (1973). *The interpretation of cultures. Basic books*. New York, NY.

Gracia-Lázaro, C., Floría, L. M., & Moreno, Y. (2011). Selective advantage of tolerant cultural traits in the Axelrod-Schelling model. *Physical Review E, 83*, 056103.

Gracia-Lázaro, C., Lafuerza, L. F., Floría, L. M., & Moreno, Y. (2009). Residential segregation and cultural dissemination: An Axelrod-Schelling model. *Physical Review E, 80*(4), 046123.

Guerra, B., Poncela, J., Gómez-Gardeñes, J., Latora, V., & Moreno, Y. (2010). Dynamical organization towards consensus in the Axelrod model on complex networks. *Physical Review E, 81*(5), 056105.

Greenberg, J. H. (1956). The measurement of linguistic diversity. *Language, 32*(1), 109–115.

Hales, S., & Hodos, T. (Eds.). (2010). *Material culture and social identities in the ancient world*. Cambridge: Cambridge University Press.

Hammond, R. A., & Axelrod, R. (2006). The evolution of ethnocentrism. *Journal of Conflict Resolution, 50*(6), 926–936.

Handwerker, P. (2008). The construct validity of cultures: Cultural diversity, culture theory, and a method for ethnography. *American Anthropologist, 104*, 106–122. https://doi.org/10.1525/aa.2002.104.1.106.

Hayden, B., & Cannon, A. (1982). The corporate group as an archaeological unit. *Journal of Anthropological Archaeology, 1*(2), 132–158.

Hirschmann, C. (2004). The origins and demise of the concept of race. *Population and Development Review, 30*(3), 385–415.

Hodder, I. (1982). *Symbols in action. Ethnoarchaeological studies of material culture*. Cambridge: Cambridge University Press.

Horn, H. S. (1966). Measurement of "overlap" in comparative ecological studies. *The American Naturalist, 100*, 419–424.

Houkes, W. (2012). Tales of tools and trees: Phylogenetic analysis and explanation in evolutionary archaeology. In *EPSA Philosophy of Science: Amsterdam 2009* (pp. 89–100). Dordrecht: Springer.

Hu, D. (2013). Approaches to the archaeology of ethnogenesis: Past and emergent perspectives. *Journal of Archaeological Research, 21*(4), 371–402.

Hudson, B. (2006). The origins of Bagan: The archaeological landscape of upper Burma to AD 1300. http://ses.library.usyd.edu.au/handle/2123/638.

Janowicz, K., Raubal, M., Schwering, A., & Kuhn, W. (2008). Semantic similarity measurement and geospatial applications. *Transactions in GIS, 12*(6), 651–659.

Jones, S. (1997). *The archaeology of ethnicity: Constructing identities in the past and present*. London: Routledge.

Kalick, S. M., & Hamilton, T. E. (1986). The matching hypothesis reexamined. *Journal of Personality and Social Psychology, 51*, 73–82.

Kandler, A., & Sherman, S. (2013). A non-equilibrium neutral model for analysing cultural change. *Journal of Theoretical Biology, 330*, 18–25. https://doi.org/10.1016/j.jtbi.2013.03.006.

Kintsch, W. (2014). Similarity as a function of semantic distance and amount of knowledge. *Psychological Review, 121*(3), 559.

Kolo, P. (2012). Measuring a new aspect of ethnicity – The appropriate diversity index, No 221, *Ibero America Institute for Econ.* Research (IAI) Discussion Papers, Ibero-America Institute for Economic Research.

Koopmans, R., & Rebers, S. (2009). Collective action in culturally similar and dissimilar groups: An experiment on parochialism, conditional cooperation, and their linkages. *Evolution and Human Behavior, 30*(3), 201–211.

Kovacevic, M., Shennan, S., Vanhaeren, M., d'Errico, F., & Thomas, M. G. (2015). Simulating geographical variation in material culture: Were early modern humans in Europe ethnically structured? In *Learning strategies and cultural evolution during the palaeolithic* (pp. 103–120). Tokyo: Springer.

Kovács, B., & Hannan, M. T. (2015). Conceptual spaces and the consequences of category spanning. *Sociological Science, 2*, 252–286.

Kim, Y. Y. (2009). The identity factor in intercultural competence. In *The Sage handbook of intercultural competence* (Vol. 1, pp. 53–65).

Larick, R. (1991). Warriors and blacksmiths: Mediating ethnicity in East African spears. *Journal of Anthropological Archaeology, 10*(4), 299–331.

Lee, J. H. (Ed.). (2016). *Morphological analysis of cultural DNA: Tools for decoding culture-embedded forms*. Singapore: Springer.

Lee, J. H. (Ed.). (2018). *Computational studies on cultural variation and heredity*. Singapore: Springer.

Lipo, C. P. (2017). The resolution of cultural phylogenies using graphs. In *Mapping our ancestors* (pp. 107–126). New York:Routledge.

Lock, G., & Molyneaux, B. L. (2006). *Confronting scale in archaeology*. New York: Springer.

Longacre, W. A., & Stark, M. T. (1992). Ceramics, kinship and space: A Kalinga example. *Journal of Anthropological Archaeology, 11*, 125–136.

Lucy, S. (2005). Ethnic and cultural identities. In M. Díaz-Andreu et al. (Eds.), *The archaeology of identity: Approaches to gender, age, status, ethnicity and religion* (pp. 86–109). London: Routledge.

Lycett, S. J. (2015). Cultural evolutionary approaches to artifact variation over time and space: Basis, progress, and prospects. *Journal of Archaeological Science, 56*, 21–31.

Madsen, M. E., & Lipo, C. P. (2014). Combinatorial structure of the deterministic seriation method with multiple subset solutions. *arXiv preprint arXiv:1412.6060.*

Madsen, M. E., & Lipo, C. (2016). Measuring Cultural Relatedness Using MultipleSeriation Ordering Algorithms. https://www.academia.edu/23561194/Measuring_Cultural_Relatedness_Using_Multiple_Seriation_Ordering_Algorithms?auto=download). Manuscript version: 2016-03-17: 3f6f16a– draft for ElectronicSymposium, "Evolutionary Archaeologies: New Approaches, Methods, And Empirical Sufficiency" at the Society for American Archaeology conference, April 2016.

Mayor, A. (2010). Ceramic traditions and ethnicity in the Niger Bend, West Africa. *Ethnoarchaeology, 2*(1), 5–48.

Mc Guire, R. H. (1982). The study of ethnicity in historical archaeology. *Journal of Anthropological Archaeology, 1*, 159–179.

Medin, D. L. (1989). Concepts and conceptual structure. *American Psychologist, 44*(12), 1469.

Medin, D. L., & Rips, L. J. (2005). Concepts and categories: Memory, meaning, and metaphysics. In *The Cambridge handbook of thinking and reasoning* (pp. 37–72). New York: Cambridge University Press.

Meskell, L. (2002). The intersections of identity and politics in archaeology. *Annual Reviews in Anthropology, 31*, 279–301.

Mills, B. J. (2017). Social network analysis in archaeology. *Annual Review of Anthropology, 46*, 379–397.

Morisita, M. (1962). Iσ-Index, a measure of dispersion of individuals. *Researches on Population Ecology, 4*(1), 1–7.

Mosterín, J. (1993). *Filosofía de la cultura*. Madrid: Alianza.

Montalvo, J. G., & Reynal-Querol, M. (2005). Ethnic polarization, potential conflict, and civil wars. *American Economic Review, 95*(3), 796–816.

Naroll, R. (1964). On ethnic unit classification. *Current Anthropology, 5*, 283–291.

Needham, R. (1975). Polythetic classification: Convergence and consequences. *Man, 10*, 349–369.

Noonan, H., & Curtis, B.(2017). Identity. In E. N. Zalta (Ed.), *The Stanford Encyclopedia of Philosophy* (Spring 2017 Edition). https://plato.stanford.edu/archives/spr2017/entries/identity/. Downloaded on 16 Apr 2018.

Norouzi, M., Fleet, D. J., & Salakhutdinov, R. R. (2012). Hamming distance metric learning. In *Advances in neural information processing systems* (pp. 1061–1069). New York: Curran Associates.

O'Brien, M., & Lyman, R. L. (2004). History and explanation in archaeology. *Anthropological Theory, 4*(2), 173–19.

O'Brien, M. J., Collard, M., Buchanan, B., & Boulanger, M. T. (2013). Trees, thickets, or something in between? Recent theoretical and empirical work in cultural phylogeny. *Israel Journal of Ecology & Evolution, 59*(2), 45–61.

O'Brien, M. J., & Lyman, R. L. (2003). *Cladistics and archaeology*. Salt Lake City: University of Utah Press.

Parker, W. S. (2015). Getting (even more) serious about similarity. *Biology and Philosophy, 30*(2), 267–276.

Peroff, N. C. (1997). Indian identity. *The Social Science Journal, 34*(4), 485–494.

Pfau, J., Kirley, M., & Kashima, Y. (2013). The co-evolution of cultures, social network communities, and agent locations in an extension of Axelrod's model of cultural dissemination. *Physica A, 392*(2), 381–391.

Pirró, G. (2009). A semantic similarity metric combining features and intrinsic information content. *Data & Knowledge Engineering, 68*(11), 1289–1308.

Pocklington, R., & Best, M. L. (1997). Cultural Evolution and units of selection in replicating text. *Journal of Theoretical Biology, 188*, 79–87.

Posner, D. (2004). Measuring ethnic fractionalization in Africa. *American Journal of Political Science, 48*(4), 849–863.

Ramasco, J. J. (2007). Social inertia and diversity in collaboration networks. *European Physical Journal Special Topics, 143*, 47–50.

Raza, G., Singh, S., & Dutt, B. (2001). Public, science, and cultural distance. *Sage Journals, 23*(3), 293–309.

Romney, A. K. (1999). Cultural consensus as a statistical model. *Current Anthropology, 40*(S1), S93–S115.

Romney, A. K., & Weller, S. C. (1984). Predicting informant accuracy from patterns of recall among individuals. *Social Networks, 4*, 59–77.

Romney, A. K., Weller, S. C., & Batchelder, W. H. (1986). Culture as consensus: A theory of culture and informant accuracy. *American Anthropologist, New Series, 88*(2), 313–338. Ed. Blackwell.

Romney, K., Boyd, J. P., Moore, C., Batchelder, W., & Brazil, T. J. (1996). Culture as shared cognitive representations. *Proceedings of the National Academy of Science, 93*, 4699–4705.

Sackett, J. R. (1990). Style and ethnicity in archaeology: The cause for Isochrestism. In M. Conkey & C. Harstof (Eds.), *The uses of style in archaeology* (pp. 32–43). New York: Cambridge University Press.

Sampson, R. J. (1984). Group size, heterogeneity, and intergroup conflict: A test of Blau's inequality and heterogeneity. *Social Forces, 62*(3), 618–639.

Santini, S., & Jain, R. (1999). Similarity measures. *IEEE Transactions on Pattern Analysis and Machine Intelligence, 21*(9), 871–883.

Schelling, T. C. (1971). Dynamic models of segregation. *Journal of Mathematical Sociology, 1*(2), 143–186.

Schudson, M. (1989). How culture works: Perspectives from media studies on the efficacy of symbols. *Theory and Society, 18*(2), 153–180.

Scott, J. (2017). *Social network analysis* (2nd ed.). London: Sage Publications.

Shenkar, O. (2001). Cultural distance revisited: Towards a more rigorous conceptualization and measurement of cultural differences. *Journal of International Business Studies, 32*(3), 519–535.

Shenkar, O. (2012). Beyond cultural distance: Switching to a friction lens in the study of cultural differences. *Journal of International Business Studies, 43*(1), 12–17.

Shennan, S. J. (Ed.). (1989). *Archaeological approaches to cultural identity*. London: Unwin Hyman.

Sieck, W. R. (2010). Cultural network analysis: Method and application. In D. Schmorrow & D. Nicholson (Eds.), *Advances in cross-cultural decision making*. Boca Raton: CRC Press/ Taylor & Francis, Ltd.

Sommer, U. (2016). Tribes, peoples, ethnicity: Archaeology and changing 'we groups'. In *Evolutionary and interpretive archaeologies* (pp. 169–198). New York:Routledge.

Sousa, C. M., & Bradley, F. (2008). Cultural distance and psychic distance: Refinements in conceptualisation and measurement. *Journal of Marketing Management, 24*(5–6), 467–488.

Squazzoni, F. (2012). *Agent-based computational sociology*. Chichester: Wiley.

Stivala, A., Robins, G., Kashima, Y., & Kirley, M. (2014). Ultrametric distribution of culture vectors in an extended Axelrod model of cultural dissemination. *Scientific Reports, 4*, 4870.

Stivala, A., Kashima, Y., & Kirley, M. (2016). Culture and cooperation in a spatial public goods game. *Physical Review E, 94*(3), 032303.

Suzuki, R., & Shimodaira, H. (2015). *Pvclust: Hierarchical clustering with P-values via multiscale bootstrap resampling*. https://CRAN.R-project.org/package=pvclust.

Tang, J., Wu, S., & Sun, J. (2013). Confluence: Conformity influence in large social networks. In *Proceedings of the 19th ACM SIGKDD international conference on Knowledge discovery and data mining* (pp. 347–355). New York: ACM. https://doi.org/10.1145/2487575.2487691.

Taylor, C. L., & Hudson, M. C. (1972). *World handbook of political and social indicators* (2nd ed.). New Haven: Yale University Press.

Tehrani, J. J. (2011). Missing links: Cultures, species and the Cladistic reconstruction of prehistory. *Evolutionary and interpretive archaeologies: A dialogue* (p. 245). New York:Routledge.

Thomas, J. (1996). *Time, culture and identity*. London: Routledge.

Thorndike, R. (1953). Who belongs in the family? *Psychometrika, 18*(4), 267–276. https://doi.org/10.1007/BF02289263.

Townley, G., Kloos, B., Green, E. P., & Franco, M. M. (2011). Reconcilable differences? Human diversity, cultural relativity, and sense of community. *American Journal of Community Psychology, 47*(1–2), 69–85.

Tversky, A. (1977). Features of similarity. *Psychological Review, 84*(4), 327.

Valori, L., Picciolo, F., Allansdottir, A., & Garlaschelli, D. (2012). Reconciling long-term cultural diversity and short-term collective social behavior. *Proceedings of the National Academy of Sciences of the United States of America, 109*, 1068–1073.

Voss, B. L. (2008). *The archaeology of ethnogenesis: Race and sexuality in colonial San Francisco*. Berkeley: University of California Press.

Wasserman, S., & Faust, K. (1994). *Social network analysis: Methods and applications* (Vol. 8). Cambridge: Cambridge University Press.

Watkins, T. (2008). Supra- regional networks in the Neolithic of Southwest Asia. *Journal of World Prehistory, 21*(2), 139–171.

Weller, S. C. (2007). Cultural consensus theory: Applications and frequently asked questions. *Field Methods, 19*(4), 339–368.

Whitmeyer, J. (1997). Endogamy as a basis for ethnic behavior. *Sociological Theory, 15*(2), 162–178.

Wilkins, J. S. (1998). What's in a meme? Reflections from the perspective of the history and philosophy of evolutionary biology. *Journal of Memetics—Evolutionary Models of Information Transmission* [Online], 2. Available: http://www.cpm.mmu.ac.uk/jom-emit/vol2/wilkins_js.html.

Williams, C. J. F. (1989). *What is identity?* Oxford: Oxford University Press.

Zhang, H. (2015). Moderate tolerance promotes tag-mediated cooperation in spatial Prisoner's dilemma game. *Physica A: Statistical Mechanics and its Applications, 424*, 52–61.

Zou, X., Tam, K. P., Morris, M. W., Lee, S. L., Lau, I. Y. M., & Chiu, C. Y. (2009). Culture as common sense: Perceived consensus versus personal beliefs as mechanisms of cultural influence. *Journal of Personality and Social Psychology, 97*(4), 579.

Modeling Niche Construction in Neolithic Europe

R. Alexander Bentley and Michael J. O'Brien

Introduction

Discussions of population size and cumulative culture (e.g., Henrich 2004; Powell et al. 2009; Shennan 2011a; Bentley and O'Brien 2012) rest on the notion that culture is adaptive over many generations, as learned traditions are slowly modified through time. This premise applies equally to such things as folk tales (Tehrani 2013) as it does to stone tools (O'Brien et al. 2001) and textiles (Tehrani and Collard 2002). In modeling ancient agricultural societies, it is essential to consider how cultural knowledge is inherited, including specific transmission pathways, often directed by kinship systems, and their feedback within small farming communities. Modeling these societies requires a rich archaeological record with excellent chronological resolution. One attractive record in this respect is the European Neolithic, dating ca. 9000–3400 years ago, because there is both a richness of the material record and a slow tempo of change, such that regional chronologies can often be resolved in considerable detail. One might, for example, observe repeated domestic practices such as wall replastering in the mudbrick houses of Çatalhöyük, in Anatolia, as often as ten times per year (Hodder and Cessford 2004) or generations of interaction between forager and farmers in various parts of Europe (Zvelebil 2006; Bollongino et al. 2013). Cultural legacies of the Neolithic, both in Europe and elsewhere, were so resilient that even millennia of rice agriculture in China apparently has left a cultural legacy in modern populations, in that rice-farming regions

R. A. Bentley (✉)
Department of Anthropology, University of Tennessee, Knoxville, TN, USA
e-mail: rabentley@utk.edu

M. J. O'Brien
Department of Arts and Humanities, Texas A&M University–San Antonio, San Antonio, TX, USA

score more holistically and collectively on social surveys than do their counterparts in regions where wheat has predominated (Talhelm et al. 2014).

In Europe, Neolithic burial practices, pottery decorations, stone tools, craft techniques, and raw material sources (Modderman 1988; Lüning et al. 1989; Gronenborn 1999) provide excellent data for testing cultural-evolutionary processes over time periods of centuries or more. As a result, the European Neolithic offers numerous insights into human adaptations to environmental change (Barnosky et al. 2012; Ehrlich and Ehrlich 2013; Hughes et al. 2013; Manning et al. 2014), agro-pastoral innovation (Bentley 2013), human-population dynamics (Shennan et al. 2013), hereditary inequality (Shennan 2011b), specialized occupations (Bentley et al. 2008), and private land ownership (Ingold 1986; Bentley et al. 2012).

The Neolithic was the first human era to leave large-scale detectable changes in the environment. Most researchers mark the Neolithic as the beginning of the Anthropocene (Foley et al. 2013; Smith and Zeder 2013), using as a hallmark a significant increase in small-scale intensive cultivation and land management (e.g., Bogucki 1993; Scarre 2000; Gartner 2001; Zimmermann et al. 2009; McMichael et al. 2012; Bogaard 2014; Lechterbeck et al. 2014; Saqalli et al. 2014). This, in turn, led to wholesale changes in the feedback between Neolithic village populations and their environments. For example, increases in atmospheric carbon dioxide and methane that began around 8000 and 5000 years ago, respectively, were likely caused by forest clearing and livestock pastoralism in Europe and early rice irrigation in East Asia (Fuller et al. 2011; Kaplan et al. 2011; Ruddiman 2013). Further, pollen evidence from about 6000 years ago suggests forest clearance by a growing population of farmers (Whitehouse et al. 2014; Woodbridge et al. 2014), such that by the beginning of the Bronze Age in Europe, about 3750 years ago, intensified land clearance had been a significant transformative force (Bradshaw 2004; Berglund et al. 2008).

Although agriculture often led to population growth as a result of increased yields, higher nutrition content, and climatic resilience of domestic plants (Bocquet-Appel 2011; Dayton 2014), we now know that on the scale of centuries, fertility rates in the Neolithic fluctuated considerably across time and place (Zimmermann et al. 2009; Shennan et al. 2013; Timpson et al. 2014). In early Neolithic Germany, for example, about 160 longhouses from the LBK culture were constructed along a one-kilometer stretch of the Merzbach River over a period of four centuries, 7300–6900 years ago (Ammerman and Cavalli-Sforza 1984), during which the initial population is estimated to have grown about 2% annually (Stehli 1989). Such a high growth, however, was both localized and offset by population declines throughout the Neolithic (Shennan et al. 2013).

Reliance on domesticated livestock also varied substantially across regions. Whether one sees uniformity versus geographic diversity in the LBK faunal record depends in large part on the chronological and spatial scales of analysis (Hofmann et al. 2012). From Early LBK settings, roughly 7500 years ago, domestic animals (Fig. 1) comprise at least 90% of faunal assemblages in the upper Rhine and upper Rhône valleys (Chaix 1997; Jeunesse and Arbogast 1997); they range from 60% to 97% in the Paris Basin (Jeunesse and Arbogast 1997) but are less than 10% at certain sites in the Alps (Chaix 1997).

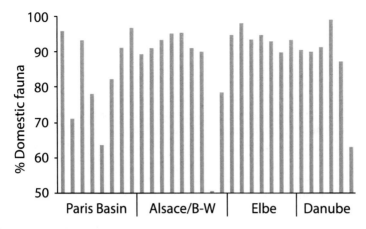

Fig. 1 Proportion of domestic fauna in France and southern Germany ca. 7500 years ago. (After Jeunesse and Arbogast 1997)

Neolithic Niche Construction

Fluctuations in the Neolithic population suggest that bottlenecks (Shennan 2000; Shennan et al. 2013) significantly affected the genetic and cultural diversity of early farming populations (Skoglund et al. 2012; Brandt et al. 2013, 2015). These bottlenecks were brought on not only by human ecological factors such as nutrition, disease resistance, and climatic adaptations (e.g., Jackes et al. 1997; Holtby et al. 2012; Bickle and Fibiger 2014) but also by social factors such as group alliances and cultural transmission, which are also fundamentally affected by population size (Henrich 2004; Bentley and O'Brien 2011). These factors spurred the intensification of *niche construction*, which, as we will see, had significant implications for the evolution of Neolithic societies.

Niche construction refers to the capacity of organisms to modify natural selection in their environment and thereby act as codirectors of their own evolution as well as that of others (Odling-Smee et al. 1996; Laland and O'Brien 2012). Under the conventional view of evolution, species, through natural selection, come to exhibit those traits, or features, that best enable them to survive and reproduce in their environments. Thus, "adaptation is always asymmetrical; organisms adapt to their environment, never vice versa" (Williams 1992: 484). Alternatively, niche construction creates adaptive symmetry by using and transforming natural selection, thus generating feedback in evolution at various levels (Laland and Sterelny 2006). To quote Levins and Lewontin (1985: 106), "the organism influences its own evolution, by being both the object of natural selection and the creator of the conditions of that selection." Niche-constructing species play important ecological roles by creating and modifying habitats and resources used by other species, thereby affecting the flow of matter and energy through ecosystems. This process, often referred to as "ecosystem engineering" (Jones et al. 1994), can have significant downstream consequences for succeeding generations, leaving behind an "ecological inheritance" (Odling-Smee 1988). As we discuss later, the modern world continues to be shaped by what it inherited from the Neolithic (Bentley et al. 2015).

An important aspect of niche construction is that acquired characters play an evolutionary role through transforming selective environments. This is particularly relevant to human evolution, where our species has engaged in extensive environmental modification through cultural practices (Laland and O'Brien 2012). This is why humans have been referred to as the "ultimate niche constructors" (Odling-Smee et al. 2003: 28). Humans can construct developmental environments that feed back to affect how individuals learn and develop and the diseases to which they are exposed. With its millennia of domestic coevolution of humans, plants, animals, and their communal landscape (Shennan 2011b; O'Brien and Laland 2012), the Neolithic witnessed the initial feedback between human diet and land management that led to changes in numerous cultural systems, including property control, wealth inheritance, and social order.

In terms of diet, faunal evidence indicates domestication of cattle, goats, and sheep began between 10,000 and 11,000 years ago (Zeder and Hesse 2000; Troy et al. 2001; Helmer et al. 2005; Peters et al. 2005). In addition to a dramatic shift in diet from Mesolithic wild animals and plants (boar, aurochs, deer, eggs, fish, shellfish, and hazelnuts) to domesticated animals and crops (wheat, barley, peas, flax, and opium poppy), the Neolithic witnessed a shift from communally shared territory for hunting and gathering to individual ownership of agricultural land (Ingold 1986), where garden plots were managed for generations by households or lineages (Bogaard 2004; Lüning 2005; Bogaard et al. 2011). The positive correlation between isotope signatures and heritable dental traits suggests that wealth and occupation had begun to be inherited, as strontium, oxygen, and carbon isotope measurements in tooth enamel, compared with heritable dental traits, reveal group structure among the victims of a Neolithic massacre (Bentley et al. 2008).

Affiliation between lineages and resources signaled the origins of hereditary socioeconomic inequality (Shennan 2011b; Bentley et al. 2012; Pringle 2014). This is seen in LBK burials, where there is evidence of ascribed status (Nieszery 1995; Jeunesse 1997), as well as in isotopic evidence from Neolithic skeletons, which shows differential access to land resources. On a continental scale, strontium-isotope measurements in teeth from over 300 Neolithic human skeletons from across Europe (Fig. 2), from eastern France to Austria, reveal that males buried with polished stone adzes had access to more local, highly productive loessial soils than those without adzes (Bentley et al. 2012).

The clear majority of non-loess isotopic signatures were found among females (Fig. 2), which supports evidence from archaeological, linguistic, and genetic studies that LBK society was largely patrilocal (Bogucki 1993; Cavalli-Sforza 1997; Eisenhauer 2003; Bentley et al. 2008, 2012; Fortunato and Jordan 2010; Bentley 2013; Brandt et al. 2015). The tendency for cattle ownership to co-occur with patrilineal kinship (Holden and Mace 2003) suggests that patriliny and livestock

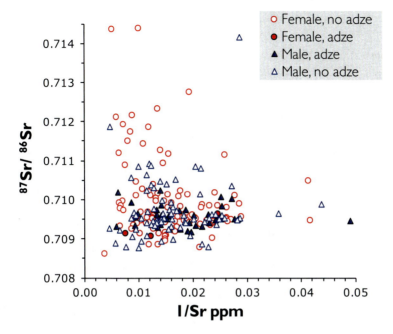

Fig. 2 Strontium isotope signatures in human enamel samples from early Neolithic sites across Central Europe reveal that males buried with polished stone adzes (filled triangles) had access to more local, highly productive loessial soils than those without adzes (open triangles). The clear majority of non-loess isotopic signatures were found among females (circles), which supports evidence from archaeological, linguistic, and genetic studies that LBK society was largely patrilocal. (After Bentley et al. 2012)

ownership would have been mutually reinforcing trends that were conducive to the growth of hereditary inequality over time, as males endeavored to retain resource access within their lineages of specialized stockkeepers and cultivators (Bogucki 1993; Eisenhauer 2003; Vigne and Helmer 2007; Bentley et al. 2008; Lazaridis et al. 2014). At the same time, women made pottery in styles learned within, and emblematic of, their matriline (Claßen 2009; Strien 2010).

This indicates that certain groups had privileged access to preferred loess soils for farming, potentially signaling the origins of hereditary social inequality (Bentley et al. 2012; Bentley 2013). At the Neolithic village of Vaihingen, in the Neckar Valley of Germany, evidence from paleobotanical remains, combined with analysis of decorative pottery motifs across groupings of houses within the village, indicates that different family lineages had access to different portions of land for their domestic animals and crops, with certain groups having access to more local, and presumably valuable, resource patches compared to other groups who had to travel farther for subsistence (Bogaard et al. 2011). It appears that different lineages within Vaihingen identified themselves through pottery decoration, craft techniques, and raw material sources (Strien 2010).

Niche Construction and the Evolution of Lactose Tolerance

The considerable biological costs of a Neolithic diet and lifestyle suggest that niche construction changed the conditions of selection on populations. Hunter-gatherers, for example, rarely exhibit obesity, diabetes, hypertension, high blood sugar, or cardiovascular disease (Cordain et al. 2002; Kaplan et al. 2017)—conditions that are now common in industrialized populations. Although diet is one obvious factor, another is the frequency and circadian timing of energy intake (Mattson et al. 2014), with hunter-gatherer diets being more intermittent. This appears to protect against severe stress by stimulating the cellular-level process of removing damaged molecules and organelles in organ and muscle tissues (Mattson et al. 2014) as well as increasing insulin sensitivity and ketone levels and reducing pro-inflammatory cytokines (Mattson et al. 2014).

In contrast, the much more regular energy intake resulting from intensive, carbohydrate-rich diets such as that of Neolithic farmers could have challenged a physiology that evolved for energy storage in liver glycogen and adipose tissue (Mattson et al. 2014). Milk, for example, contains lactose, a disaccharide that must be broken down by the enzyme lactase into the two monosaccharides glucose and galactose to be digested. Without lactase, drinking milk leads to a battery of symptoms, including diarrhea, cramping, gas, nausea, and vomiting. For most modern populations, with access to medical care if needed, this is not much of a problem, but for those without access, untreated diarrhea can be fatal. Prior to the Neolithic, adults had no ability to produce lactase.

If dairy products can be digested, however, they become an excellent supply of fat, proteins, carbohydrates, vitamins, calcium, and water (Wooding 2007). Hence, Neolithic individuals with the biological capacity for adult lactose tolerance and carbohydrate digestion would have had a selective advantage, increasing lactose-tolerant lineages in a population over time (Itan et al. 2009, 2010; Laland et al. 2010; Gerbault et al. 2011; O'Brien and Laland 2012; O'Brien and Bentley 2015). Evidence from both faunal remains and milk residues on pottery indicate that milk production began about 8000 years ago in Mediterranean Europe and the Middle East (Vigne and Helmer 2007; Conolly et al. 2011, 2012; Manning et al. 2014), 7000–8000 years ago in northwestern Anatolia (Dudd and Evershed 1998), 7000 years ago in central Europe (Salque et al. 2013), and about 6000 years ago in Britain (Copley et al. 2003, 2005; Evershed et al. 2008; Craig 2011). We assume European LP and dairying began to slowly coevolve between ca. 6250 and 8700 years ago, somewhere between the northern Balkans and central Europe (Itan et al. 2009), followed by a rapid demographic expansion into central and north-central Europe by way of the LBK culture 7500–8000 years ago (Dolukhanov et al. 2005; Edwards et al. 2007), after which cattle-based dairying economies established themselves about 6500 years ago.

As an instance of apparent niche construction, Neolithic dairy farming and cereal cultivation led to significant changes in the digestive system of a number of populations. In Europe, Neolithic populations evolved a single nucleotide polymorphism

(−*13910*T*) that allowed them and their descendants to digest milk (Enattah et al. 2002). Whether or not the lactase-persistent (LP) allele achieved high frequency depended on the probability of the offspring of milk drinkers becoming milk drinkers themselves (Feldman and Cavalli-Sforzi 1989). In other words, milk drinking had to be not just a learned tradition but a *reliably* learned tradition (Aoki 1996). Once milk drinking became a tradition that was learned consistently enough from one generation to the next, a significant fitness advantage to individuals with the LP allele fixed it within a few 1000 years—one of the strongest known evolutionary cases of selection for a human allele (Bersaglieri et al. 2004).

If certain groups inherited livestock herding as their lineage specialization (Bogucki 1993; Bentley et al. 2008; Bentley 2013), then wealthy, lactose-tolerant, cattle-owning lineages may have had a selective advantage, especially during population bottlenecks. Cattle ownership in the Neolithic surely signified both wealth and/or status (Bogucki 1993), with owners of large dairy herds having an advantage over smaller herd owners or nonowners and tending to absorb the smaller herds after a crisis (Salzman 1999; Hayden 2001; Zvelebil 2006). If climatic adversity undermined the stability of Neolithic societies (Gronenborn 2007), then wealth was a likely factor in long-term survival of lineages in the Neolithic, as livestock wealth often predicts an increase in reproductive success (Holden and Mace 2003; Alesina et al. 2011; Sear 2015).

O'Brien and Laland (2012) made these feedbacks explicit by devising a path diagram for the evolution of LP (Fig. 3). The figure represents how domestication of cattle triggers (1) milk consumption, which (2) favors the spread of LP, which (3) promotes further milk consumption, which (4) elicits further milk-product manufacture and consumption, which (5) leads to selective breeding of cattle, and which (6) selects for alleles conferring high milk yield in dairy cattle. In addition, cattle

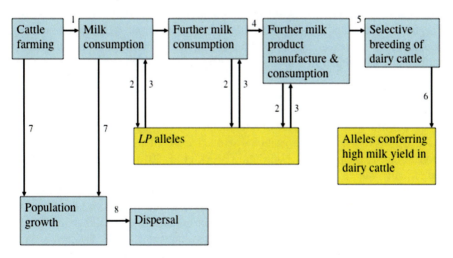

Fig. 3 Path diagram for the niche-construction explanation for the evolution of lactase persistence. (After O'Brien and Laland 2012)

farming and dairy product consumption (7) lead to population growth, which (8) triggers dispersal into new environments.

The variables in Fig. 3 can be labeled as follows:

- C_t Dairy-animal husbandry
- N_t Population
- M_t Milk-product consumption
- LP_t LP alleles
- FM_t Further consumption of milk products
- FMP_t Further milk-product manufacture and consumption
- SB_t Selective breeding of dairy animals
- AM_t Alleles conferring high milk yield in dairy animals
- D_t Dispersal at date t

Testing for the Presence of Neolithic Niche Construction

Matthews et al. (2014) describe a set of criteria to test for the presence of niche construction (criteria 1 and 2) and to determine when it affects evolution (criterion 3). In simplified terms, these are the following: (1) an organism must significantly modify environmental conditions; (2) organism-mediated environmental modifications must influence selection pressures on a recipient organism; and (3) there must be an evolutionary response in at least one recipient population caused by the environmental modification.

As Laland et al. (2016) note, criteria 1 and 2 are sufficient for the definition of niche construction, but it is the all-important third criterion that is the test of evolution by niche construction. As they further note, implementation of criterion 1 depends on how the term "significantly" is interpreted: "Whether a form of environmental modification is recognized as 'niche construction' requires a pragmatic judgment by the researcher as to whether . . . the environmental modification is sufficiently substantial in scale, duration and impact to plausibly affect selection" (Laland et al. 2016: 193).

With respect to our case example, O'Brien and Laland (2012) proposed that the coevolution of Neolithic dairying and the LP allele in Europe was a classic case of niche construction—a perspective also adopted by Gerbault et al. (2011). And, the case appears to meet criteria 1 and 2 discussed above. It also appears to meet criterion 3, but as we point out elsewhere (Brock et al. 2016), demonstrating that fit empirically can be a difficult enterprise, especially given the number of potential feedbacks involved (Brock et al. 2016). In short, do we have the data to demonstrate empirically which changes to which variable(s) occurred first? Do we have the data to support the directionality of various time series, such as those shown in Fig. 3?

To begin to examine temporal relationships among any set of variables, we can use Granger (1969) causality, which is a linear modeling approach that generates statements about the incremental predictive value of time series (Geweke 1984;

Brock et al. 2016). Under this framework, time series D is said to "Granger cause" time series U if past values of both D and U incrementally predict future values of U better than past values of U alone. Note what this statement does *not* say: It does not say that time series D *causes* time series U. Rather, it says that, based on the evidence, time series D "Granger causes" time series U. Granger causality uses the following bivariate linear autoregressive model:

$$U(t) = \sum_{j=1}^{p} A_{11,j} U(t-j) + \sum_{j=1}^{p} A_{12,j} D(t-j) + \varepsilon_1(t)$$
$$D(t) = \sum_{j=1}^{p} A_{21,j} U(t-j) + \sum_{j=1}^{p} A_{22,j} D(t-j) + \varepsilon_2(t)$$
(1)

where p is the maximum number of lagged observations in the model, A is a matrix of coefficients, $U(t)$ is the dependent variable, and $D(t)$ is the Granger causal variable. The regression errors ε_1 and ε_2 are assumed to be conditionally independent of the regressors at each time interval t (Greene 2003).

In their analysis of Neolithic dairying as an example of niche construction, Brock et al. (2016) considered the linear autoregressive system below:

$$M_{t+1} = \alpha_{10} + \alpha_{11} M_t + \alpha_{12} \mathrm{LP}_t + e_{1,t+1}, t = 1, 2, \ldots$$
$$\mathrm{LP}_{t+1} = \alpha_{20} + \alpha_{21} M_t + \alpha_{22} \mathrm{LP}_t + e_{2,t+1}, t = 1, 2, \ldots$$
(2)

There are just two variables in this system: LP denotes fractions of people with the LP allele in the gene pool, and M denotes the number of dairy animals. This formulation allows only direct influences between LP and M (via coefficients α_{ij}) rather than the indirect influences of domestication as the start of the human-constructed niche (O'Brien and Laland 2012).

We assume the system (2) is stationary as well as initialized with a very small value of LP_1. In this simple formulation (e.g., Scott-Phillips et al. 2014), domestication is the "environment," and the density of dairy animals at date t C_t, unidirectionally "causes" the frequency of lactase persistence at date $t+1$, LP_{t+1}. With niche construction, however, we would expect that Neolithic groups with the LP allele at date t affected the number of dairy animals at period $t+1$ above and beyond the mere effect of the number of people at date t:

$$M_{t+1} = \alpha_{10} + \alpha_{11} M_t + \alpha_{12} \mathrm{LP}_t + \alpha_{13} N_t + e_{1,t+1}$$
$$\mathrm{LP}_{t+1} = \alpha_{20} + \alpha_{21} M_t + \alpha_{22} \mathrm{LP}_t + \alpha_{23} N_t + e_{2,t+1}$$
$$N_{t+1} = \alpha_{30} + \alpha_{31} M_t + \alpha_{32} \mathrm{LP}_t + \alpha_{33} N_t + e_{3,t+1}$$
(3)

In systems (2) and (3), the frequency of alleles governs a population's capacity for niche construction (Laland et al. 1999; Han and Hui 2014). In order to estimate the path-link strengths, $\{a_{ij}, i = 1, 2, 3, j = 1, 2, 3\}$, we need empirical data representative of multiple time periods, $t = 1, 2\ldots$, and enough dated measures for M_t, LP_t, N_t so that we can estimate the parameters of Eq. (3) using regression analysis. Brock et al. (2016) translate the path diagram in Fig. 1 by using the linear-dynamic-system assumption of Eqs. (1) and (2) but with more variables and equations, such that each path can be tested by ordinary least squares (OLS).

Brock et al. (2016) translated the path diagram in Fig. 3 into a series of coupled equations as

$$\begin{aligned} M_{t+1} &= a_{10} + b_{12}C_t + a_{13}\text{LP}_t + a_{12}N_t + e_{1t+1} \\ N_{t+1} &= a_{20} + b_{22}C_t + a_{21}M_t + a_{22}N_t + e_{2,t+1} \\ \text{LP}_{t+1} &= a_{30} + b_{32}C_t + a_{31}M_t + a_{33}\text{LP}_t + a_{34}\text{FM}_t + a_{35}\text{FMP}_t + e_{3,t+1} \\ \text{FM}_{t+1} &= a_{40} + a_{41}M_t + a_{43}\text{LP}_t + a_{42}N_t + e_{4,t+1} \\ \text{FMP}_{t+1} &= a_{50} + a_{53}\text{LP}_t + a_{54}\text{FM}_t + a_{52}N_t + e_{5,t+1} \\ \text{SB}_{t+1} &= a_{60} + a_{65}\text{FMP}_t + a_{62}N_t + e_{6,t+1} \\ \text{AM}_{t+1} &= a_{70} + a_{76}\text{SB}_t + e_{7,t+1} \\ D_{t+1} &= a_{80} + a_{82}N_t + e_{8,t+1} \end{aligned} \qquad (4)$$

The series of Equations (4) can be represented in vector-matrix form as follows:

$$X_{t+1} = a_0 + AX_t + b \otimes C_t + e_{t+1}, \qquad (5)$$

where X is an 8×1 column vector, A is an 8×8 matrix, C_t is the number of dairy animals at time t (a scalar), b is an 8×1 column vector, and the Kronecker product $b \otimes C_t$ is an 8×1 vector. Equation (5) describes an auto-regressive system with one lag (Brock et al. 2016). If we assume the square matrix A is stable, i.e., the errors are independent of the regressors at each date t, we may estimate each equation in (4) by OLS, equation by equation, yielding estimates of the parameters of matrix A (Brock et al. 2016).

What data can we fit to these equations? We start by assuming that we have data sources for several of these proxies but no data for FM_t, FMP_t, SB_t, and AM_t because these cannot (yet, anyway) be observed directly. How well we can estimate the parameters will depend, of course, on the temporal resolution of each variable. Unfortunately, each time series may contain only one point per millennium or even fewer. Some data are also indirect, such as those used to estimate Neolithic population densities. The best proxy for N_t may be the 8000 radiocarbon dates from across western Europe that Shennan et al. (2013) used as a proxy for population size (Fig. 4), with radiocarbon dates counted in each region and binned by time interval—about one bin for every 16 years—with a correction for an assumed exponential decay in probability of site discovery with age.

There are demographic data from Neolithic skeletal assemblages, but their temporal resolution is poor. Bocquet-Appel (2011) aggregated data from 133 cemeteries during the transition from foraging to farming, with 23 pre-Neolithic datum points spread over 4000 years, or about one every 160 years, and 100 Neolithic and post-Neolithic datum points spread over the subsequent 4000 years, or about one every 40 years. Similarly for the dispersal variable, D_t, the data sets assembled are impressive

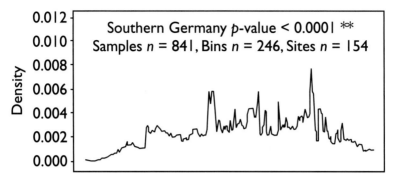

Fig. 4 Radiocarbon dates from southern Germany, with 841 dates binned into 246 time slices over a period of about 4000 years. (After Shennan et al. 2013)

(e.g., Gkiasta et al. 2003; Pinhasi et al. 2005), but when binned into temporal units, we arrive at time series comprising one or two datum points per millennium.

Let's move to the next variable, the density of domestic animals, C_t, and consider one of the most comprehensive studies to date, covering 400,000 animal bones recovered from 114 archaeological sites in southeastern Europe and Southwest Asia (Conolly et al. 2011, 2012). Even with such a massive body of faunal evidence, once the minimum-number-of-individuals counts have been assigned to temporal bins between the 12th and 8th millennia before present, we are left with about one datum point per millennium. Other estimates of C_t might be derived theoretically, such as the optimal foraging model estimating a typical six-household Neolithic village as using six square kilometers of land for cultivation and grazing a herd of about 40 cattle and 40 sheep/goats (Gregg 1988). The problem, of course, is that this is still just one datum point in the time series, which not only took a considerable amount of detailed research to compile but was focused on only one specific region of southwest Germany (Gregg 1988).

When subdivided over geography, there are even fewer counts per bin, such that the aggregation on a continental scale masks the volatility of change over time. In estimating Neolithic population, N_t, for example, local growth rates may have risen well above the continental mean for short periods, as described above. Similarly, the number of cattle, C_t, might have varied substantially from one region to another due to differences in overall adaptive subsistence or in the specializations within food production exchange systems.

Conclusion

We applaud our colleagues who have embraced niche construction as an important process in human biological and cultural evolution (e.g., Smith 2007, 2012, 2016; Kendal et al. 2011; Shennan 2011b; Boivin et al. 2016; Zeder 2016) and hope that our brief discussion of the European Neolithic adds to the conversation. Our point is that it can be easy to invoke niche construction but difficult to test it empirically. As we have suggested, a method such as Granger causality can be used to resolve various selection pathways that niche construction can take. As a cautionary note—and one that will come as no surprise to anyone—we point out that although niche construction might be invoked when dealing with certain cases of domestication and dairying, it would be the height of fallacy to use one case study as a proxy for another. For example, it is now clear that in Mongolia, dairying was practiced for some 4000 years prior to the appearance of the adult ability to digest lactose (Jeong et al. 2018). In modern Mongolia, herders receive more than a third of their calories from dairy products, yet 95% of them are lactose intolerant (Curry 2018). The trick is that they use bacteria to digest the lactose, which renders milk into products such as cheese and yogurt. This also occurred in the early European Neolithic, but there, the cultural push toward dairy-product consumption was met with a genetic mutation that spurred the growth of dairy-product consumption in the form of milk (Itan et al. 2009; Gerbault et al. 2011).

Changes in diet during the Neolithic reflect feedback loops within evolutionary niches that favored new means of obtaining and processing food and genetic changes for phenotypes that benefitted nutritionally from them (Laland et al. 2010; Shennan 2011b). It is ironic that although Neolithic agriculture supported larger populations through more intensive land use, it also led to a decrease in human stature, an increase in disease load on populations, and poorer nutrition compared to hunter-gatherer diets. Today, we continue to face the results of Neolithic niche construction. Modern heart disease, for example, arises from a complex interaction of the domesticated foods inherited from Neolithic ancestors—dairy products, cereals, sugars, fatty meats, and salt. Neolithic foods, in turn, adversely influence the major nutritional factors of chronic diseases of Western civilization, including glycemic load, fatty acid composition, acid-base balance, sodium/potassium ratio, and fiber content. Hence, understanding contemporary "diseases of affluence" requires a deeper understanding of the intersection between our gene-culture evolutionary history with the nutritional qualities of recently introduced foods (Cordain et al. 2005).

Although the testing of niche construction in the European Neolithic and elsewhere presents a steep challenge in terms of data, the effort to generate and aggregate data will be worth it in terms of what they tell us about rapid cultural and biological evolution of prehistoric societies. Selective feedback from human cultural activities to human genes, as well as to those of other species, may be a general feature of human evolution. Now that geneticists have identified several 100 human genes that appear to have subject to selective sweeps over the last 50,000 years or less, it may be that the coevolution of genes and culture is perhaps the dominant form of human evolution (Feldman and Laland 1996; Laland et al. 2010; Richerson et al. 2010). If so, then there is all the more reason to adopt the kind of analytical framework discussed briefly here. With respect to the Neolithic in particular, its continuing relevance has never been more apparent (Bentley et al. 2015).

Acknowledgments We thank Mehdi Saqalli and Marc Vander Linden for their kind invitation to contribute to this volume.

References

Alesina, A., Giuliano, P., & Nunn, N. (2011). Fertility and the plough. *American Economic Review, 101*, 499–503.

Ammerman, A. J., & Cavalli-Sforza, L. L. (1984). *The Neolithic transition and the genetics of populations in Europe.* Princeton: Princeton University Press.

Aoki, K. A. (1996). Stochastic model of gene–culture coevolution suggested by the "culture historical hypothesis" for the evolution of adult lactose absorption in humans. *Proceedings of the National Academy of Sciences, 83*, 2929–2933.

Barnosky, A. D., Hadly, E. A., Bascompte, J., Berlow, E. L., Brown, J. H., Fortelius, M., Getz, W. M., et al. (2012). Approaching a state shift in Earth's biosphere. *Nature, 486*, 52–58.

Bentley, R. A. (2013). Mobility and the diversity of early Neolithic lives: Isotopic evidence from skeletons. *Journal of Anthropological Archaeology, 32*, 303–312.

Bentley, R. A., & O'Brien, M. J. (2011). The selectivity of cultural learning and the tempo of cultural evolution. *Journal of Evolutionary Psychology, 9*, 125–141.

Bentley, R. A., & O'Brien, M. J. (2012). Cultural evolutionary tipping points in the storage and transmission of information. *Frontiers in Psychology, 3*, 1–14.

Bentley, R. A., Wahl, J., Price, T. D., & Atkinson, T. C. (2008). Isotopic signatures and hereditary traits: Snapshot of a Neolithic community in Germany. *Antiquity, 82*, 290–304.

Bentley, R. A., Bickle, P., Fibiger, L., Nowell, G. M., Dale, C. W., Hedges, R. E. M., Hamilton, J., et al. (2012). Community differentiation and kinship among Europe's first farmers. *Proceedings of the National Academy of Sciences, 109*, 9326–9330.

Bentley, R. A., O'Brien, M. J., Manning, K., & Shennan, S. (2015). On the relevance of the European Neolithic. *Antiquity, 89*, 1203–1210.

Berglund, B. E., Persson, T., & Björkman, L. (2008). Late Quaternary landscape and vegetation diversity in a North European perspective. *Quaternary International, 184*, 187–194.

Bersaglieri, T., Sabeti, P. C., Patterson, N., Vanderploeg, T., Schaffner, S. F., Drake, J. A., Rhodes, M., et al. (2004). Genetic signatures of strong recent positive selection at the lactase gene. *American Journal of Human Genetics, 74*, 1111–1120.

Bickle, P., & Fibiger, L. (2014). Ageing, childhood, and social identity in the early Neolithic of Central Europe. *European Journal of Archaeology, 17*, 208–228.

Bocquet-Appel, J. P. (2011). When the world's population took off: The springboard of the Neolithic demographic transition. *Science, 333*, 560–561.

Bogaard, A. (2004). *Neolithic farming in Central Europe.* London: Routledge.

Bogaard, A. (2014). Framing farming: A multi-stranded approach to early agricultural practice in Europe. In A. Whittle & P. Bickle (Eds.), *Early farmers: The view from archaeology and science* (pp. 181–196). Oxford: Oxford University Press.

Bogaard, A., Strien, H.-C., & Krause, R. (2011). Towards a social geography of cultivation and plant use in an early farming community: Vaihingen an der Enz, south-west Germany. *Antiquity, 85*, 395–416.

Bogucki, P. (1993). Animal traction and household economies in Neolithic Europe. *Antiquity, 67*, 492–503.

Boivin, N. L., Zeder, M. A., Fuller, D. Q., Crowther, A., Larson, G., Erlandson, J. M., Denham, T., & Petraglia, M. D. (2016). Ecological consequences of human niche construction: Examining long-term anthropogenic shaping of global species distributions. *Proceedings of the National Academy of Sciences, 113*, 6388–6396.

Bollongino, R., Nehlich, O., Richards, M. P., Orschiedt, J., Thomas, M. G., Sell, C., Fajkošová, Z., et al. (2013). 2000 years of parallel societies in Stone Age Central Europe. *Science, 342*, 479–481.

Bradshaw, R. H. (2004). Past anthropogenic influence on European forests and some possible genetic consequences. *Forest Ecology and Management, 197*, 203–212.

Brandt, G., Haak, W., Adler, C. J., Roth, C., Szécsényi-Nagy, A., Karimnia, S., Möller-Rieker, S., et al. (2013). Ancient DNA reveals key stages in the formation of central European mitochondrial genetic diversity. *Science, 342*, 257–261.

Brandt, G., Szécsényi-Nagy, A., Roth, C., Alt, K. W., & Haak, W. (2015). Human paleogenetics of Europe—The known knowns and the known unknowns. *Journal of Human Evolution, 79*, 73–92.

Brock, W. A., O'Brien, M. J., & Bentley, R. A. (2016). Validating niche-construction theory through path analysis. *Archaeological and Anthropological Sciences, 8*, 819–837.

Cavalli-Sforza, L. L. (1997). Genetic and cultural diversity in Europe. *Journal of Anthropological Research, 53*, 383–404.

Chaix, L. (1997). La transition Méso-Néolithique quelques donnés de l'archéozoologie dans les Alpes du Nord et ae Jura. In C. Jeunesse (Ed.), *Le Néolithique Danubien et ses Marges entre Rhin et Seine* (pp. 191–196). Alsace-Lorraine: Cahiers de Association pour la Promotion de la Recherche Archeologique en Alsace.

Claßen, E. (2009). Settlement history, land use and social networks of early Neolithic communities in western Germany. In D. Hofmann & P. Bickle (Eds.), *Creating communities: New advances in Central European Neolithic Research* (pp. 95–110). Oxford: Oxbow Books.

Conolly, J., Colledge, S., Dobney, K., Vigne, J.-D., Peters, J., Stopp, B., Manning, K., & Shennan, S. (2011). Meta-analysis of zooarchaeological data from SW Asia and SE Europe provides insight into the origins and spread of animal husbandry. *Journal of Archaeological Science, 38*, 538–545.

Conolly, J., Manning, K., Colledge, S., Dobney, K., Shennan, S., & S. (2012). Species distribution modeling of ancient cattle from early Neolithic sites in SW Asia and Europe. *Holocene, 22*, 997–1010.

Copley, M. S., Berstan, R., Dudd, S. N., Docherty, G., Mukherjee, A. J., Straker, V., Payne, S., & Evershed, R. P. (2003). Direct chemical evidence for widespread dairying in prehistoric Britain. *Proceedings of the National Academy of Sciences, 100*, 1524–1529.

Copley, M. S., Berstan, R., Dudd, S. N., Aillaud, S., Mukherjee, A. J., Straker, V., Payne, S., & Evershed, R. P. (2005). Processing of milk products in pottery vessels through British prehistory. *Antiquity, 79*, 895–908.

Cordain, L., Eaton, S. B., Miller, J. B., Mann, N., & Hill, K. (2002). The paradoxical nature of hunter–gatherer diets: Meat-based, yet non-atherogenic. *European Journal of Clinical Nutrition, 56*, S42–S52.

Cordain, L., Eaton, S. B., Sebastian, A., Mann, N., Lindeberg, S., Watkins, B. A., O'Keefe, J. H., & Brand-Miller, J. (2005). Origins and evolution of the Western diet: Health implications for the 21st century. *American Journal of Clinical Nutrition, 81*, 341–354.

Craig, O. E. (2011). The development of dairying in Europe: Potential evidence from food residues on ceramics. *Documenta Praehistorica, 29*, 97–107.

Curry, A. (2018). Early Mongolians ate dairy, but lacked the gene to digest it. *Science, 362*, 626–627.

Dayton, L. (2014). Blue-sky rice. *Nature, 514*, S52–S54.

Dolukhanov, P., Shukurov, A., Gronenborn, D., Sokoloff, D., Timofeev, V., & Zaitseva, G. (2005). The chronology of Neolithic dispersal in Central and Eastern Europe. *Journal of Archaeological Science, 32*, 1441–1458.

Dudd, S. N., & Evershed, R. P. (1998). Direct demonstration of milk as an element of archaeological economies. *Science, 282*, 1478–1481.

Edwards, C. J., Bollongino, R., Scheu, A., Chamberlain, A., Tresset, A., Vigne, J. D., & Baird, J. F. (2007). Mitochondrial DNA analysis shows a Near Eastern Neolithic origin for domestic cattle and no indication of domestication of European aurochs. *Proceedings of the Royal Society B, 274*, 1377–1385.

Ehrlich, P. R., & Ehrlich, A. H. (2013). Can a collapse of global civilization be avoided? *Proceedings of the Royal Society B, 280*, 20122845.

Eisenhauer, U. (2003). Jüngerbandkeramische residenzregeln: Patrilokalität in Talheim. In J. Eckert, U. Eisenhauer, & A. Zimmermann (Eds.), *Archäologische Perspektiven: Analysen und Interpretationen im Wandel* (pp. 561–573). Rahden: Leidorf.

Enattah, N. S., Sahi, T., Savilahti, E., Terwilliger, J. D., Peltonen, L., & Järvelä, I. (2002). Identification of a variant associated with adult-type hypolactasia. *Nature Genetics, 30*, 233–237.

Evershed, R. P., Payne, S., Sherratt, A. G., Copley, M. S., Coolidge, J., Urem-Kotsu, D., Kotsakis, K., et al. (2008). Earliest date for milk use in the Near East and southeastern Europe linked to cattle herding. *Nature, 455*, 528–531.

Feldman, M. W., & Cavalli-Sforzi, L. L. (1989). On the theory of evolution under genetic and cultural transmission with application to the lactose absorption problem. In M. W. Feldman (Ed.), *Mathematical evolutionary theory* (pp. 145–173). Princeton: Princeton University Press.

Feldman, M. W., & Laland, K. N. (1996). Gene–culture coevolutionary theory. *Trends in Ecology & Evolution, 11*, 453–457.

Foley, S. F., Gronenborn, D., Andreae, M. O., Kadereit, J. W., Esper, J., Scholz, D., Pöschl, U., et al. (2013). The Palaeoanthropocene—The beginnings of anthropogenic environmental change. *Anthropocene, 3*, 83–88.

Fortunato, L., & Jordan, F. (2010). Your place or mine? A phylogenetic comparative analysis of marital residence in Indo-European and Austronesian societies. *Philosophical Transactions of the Royal Society B, 365*, 3913–3922.

Fuller, D., Van Etten, J., Manning, K., Castillo, C., Kingwell-Banham, E., Weisskopf, A., Qin, L., et al. (2011). The contribution of rice agriculture and livestock pastoralism to prehistoric methane levels: An archaeological assessment. *Holocene, 21*, 743–759.

Gartner, W. G. (2001). Late Woodland landscapes of Wisconsin: Ridged fields, effigy mounds and territoriality. *Antiquity, 73*, 671–683.

Gerbault, P., Liebert, A., Itan, Y., Powell, A., Currat, M., Burger, J., Swallow, D. M., & Thomas, M. G. (2011). Evolution of lactase persistence: An example of human niche construction. *Philosophical Transactions of the Royal Society B, 366*, 863–877.

Geweke, J. (1984). Inference and causality in economic time series. In Z. Griliches & M. D. Intriligator (Eds.), *Handbook of econometrics* (Vol. 2, pp. 1101–1144). Amsterdam: North-Holland.

Gkiasta, M., Russell, T., Shennan, S. J., & Steele, J. (2003). Neolithic transition in Europe: The radiocarbon record revisited. *Antiquity, 77*, 45–62.

Granger, C. W. J. (1969). Investigating causal relations by econometric models and cross-spectral methods. *Econometrica, 37*, 424–438.

Greene, W. H. (2003). *Econometric analysis* (5th ed.). Upper Saddle River: Prentice Hall.

Gregg, S. (1988). *Foragers and farmers: Population interaction and agricultural expansion in prehistoric Europe*. Chicago: University of Chicago Press.

Gronenborn, D. (1999). A variation on a basic theme: The transition to farming in southern Central Europe. *Journal of World Prehistory, 13*, 123–210.

Gronenborn, D. (2007). Beyond the models: 'Neolithisation' in Central Europe. *Proceedings of the British Academy, 144*, 73–98.

Han, X., & Hui, C. (2014). Niche construction on environmental gradients: The formation of fitness valley and stratified genotypic distributions. *PLoS One, 9*(6), e99775.

Hayden, B. (2001). The dynamics of wealth and poverty in the transegalitarian societies of Southeast Asia. *Antiquity, 75*, 571–581.

Helmer, D., Gourichon, L., Monchot, H., Peters, J., & Sana Segui, M. (2005). Identifying early domestic cattle from pre-pottery Neolithic sites on the Middle Euphrates using sexual dimorphism. In J. D. Vigne, J. Peters, & D. Helmer (Eds.), *The first steps of animal domestication: New archaeozoological approaches* (pp. 86–95). Oxford: Oxbow.

Henrich, J. (2004). Demography and cultural evolution: Why adaptive cultural processes produced maladaptive losses in Tasmania. *American Antiquity, 69*, 197–214.

Hodder, I., & Cessford, C. (2004). Daily practice and social memory at Çatalhöyük. *American Antiquity, 69*, 17–40.

Hofmann, D., Bentley, R. A., Bickle, P., Bogaard, A., Crowther, J., Cullen, P., Fibiger, L., et al. (2012). Kinds of diversity and scales of analysis in the LBK. In F. Kreienbrink, M. Cladders, H. Stäuble, T. Tischendorf, & S. Wolfram (Eds.), *Siedlungsstruktur und Kulturwandel in der Banderkermik* (pp. 103–113). Dresden: Landesamt fur Archäologie.

Holden, C. J., & Mace, R. (2003). Spread of cattle led to the loss of matrilineal descent in Africa: A co-evolutionary analysis. *Proceedings of the Royal Society B, 270*, 2425–2433.

Holtby, I., Scarre, C., Bentley, R. A., & Rowley-Conwy, P. (2012). Disease, CCR5-32 and the European spread of agriculture? A hypothesis. *Antiquity, 86*, 207–210.

Hughes, T. P., Carpenter, S., Rockström, J., Scheffer, M., & Walker, B. (2013). Multiscale regime shifts and planetary boundaries. *Trends in Ecology & Evolution, 28*, 389–395.

Ingold, T. (1986). *The appropriation of nature: Essays on human ecology and social relations*. Manchester: Manchester University Press.

Itan, Y., Powell, A., Beaumont, M. A., Burger, J., & Thomas, M. G. (2009). The origins of lactase persistence in Europe. *PLoS Computational Biology, 5*(8), e1000491.

Itan, Y., Jones, B. L., Ingram, C. J., Swallow, D. M., & Thomas, M. G. (2010). A worldwide correlation of lactase persistence phenotype and genotypes. *BMC Evolutionary Biology, 10*, 36.

Jackes, M., Lubell, D., & Meiklejohn, C. (1997). Healthy but mortal: Human biology and the first farmers of Western Europe. *Antiquity, 71*, 639–658.

Jeong, C., Wilkin, S., Amgalantugs, T., Bouwman, A. S., Taylor, W. T. T., Hagan, R. W., Bromage, S., et al. (2018). Bronze Age population dynamics and the rise of dairy pastoralism on the eastern Eurasian steppe. *Proceedings of the National Academy of Sciences, 115*, E11248–E11255.

Jeunesse, C. (1997). *Pratiques Funéraires au Néolithique Ancien. Sépultures Nécropoles Danubiennes 5500–4900 av. J.-C*. Paris: Éditions Errance.

Jeunesse, C., & Arbogast, R. M. (1997). A propos du statut de la chasse au Neolithique moyen. La faune sauvage dans les dechets domestiques et dans les mobiliers funeraires. In C. Jeunesse (Ed.), *Le Néolithique Danubien et ses Marges entre Rhin et Seine* (pp. 81–102). Alsace-Lorraine: Cahiers de Association pour la Promotion de la Recherche Archeologique en Alsace.

Jones, C. G., Lawton, G. H., & Shachak, M. (1994). Organisms as ecosystem engineers. *Oikos, 69*, 373–386.

Kaplan, J. O., Ellis, E. C., Ruddiman, W. F., Lemmen, C., & Goldewijk, K. K. (2011). Holocene carbon emissions as a result of anthropogenic land cover change. *Holocene, 21*, 775–791.

Kaplan, H., Thompson, R. C., Trumble, B. C., Wann, L. S., Allam, A. H., Beheim, B., Frohlich, B., et al. (2017). Coronary atherosclerosis in indigenous South American Tsimane: A cross-sectional cohort study. *Lancet, 389*, 1730–1739.

Kendal, J. R., Tehrani, J. J., & Odling-Smee, F. J. (2011). Human niche construction in interdisciplinary focus. *Philosophical Transactions of the Royal Society B, 366*, 785–792.

Laland, K. N., & O'Brien, M. J. (2012). Cultural niche construction: An introduction. *Biological Theory, 6*, 191–202.

Laland, K. N., & Sterelny, K. (2006). Seven reasons (not) to neglect niche construction. *Evolution, 60*, 1751–1762.

Laland, K. N., Odling-Smee, F. J., & Feldman, M. W. (1999). The evolutionary consequences of niche construction and their implications for ecology. *Proceedings of the National Academy of Sciences, 96*, 10242–10247.

Laland, K. N., Odling-Smee, J., & Myles, S. (2010). How culture shaped the human genome. *Nature Reviews Genetics, 11*, 137–148.

Laland, K. N., Matthews, B., & Feldman, M. W. (2016). An introduction to niche construction theory. *Evolutionary Ecology, 30*, 191–202.

Lazaridis, I., Patterson, N., Mittnik, A., Renaud, G., Mallick, S., Kirsanow, K., Sudmant, P. H., et al. (2014). Ancient human genomes suggest three ancestral populations for present-day Europeans. *Nature, 513*, 409–413.

Lechterbeck, J., Edinborough, K., Kerig, T., Fyfe, R., Roberts, N., & Shennan, S. (2014). Is Neolithic land use correlated with demography? *Holocene, 24*, 1297–1307.

Levins, R., & Lewontin, R. C. (1985). *The dialectical biologist*. Cambridge, MA: Harvard University Press.

Lüning, J. (2005). Bandkeramische Hofplätze und absolute chronologie der Bandkeramik. In J. Lüning, C. Fridrich, & A. Zimmerman (Eds.), *Die Bandkeramik im 21 Jahrhundert* (pp. 49–74). Rahden: Leidorf.

Lüning, J., Kloos, U., & Albert, S. (1989). Westliche Nachbarn der Bandkeramischen Kultur: La Hoguette und Limburg. *Germania, 67*, 355–393.

Manning, K. M., Downey, S. S., Colledge, S., Conolly, J., Stopp, B., Dobney, K., & Shennan, S. (2014). The origins and spread of stock-keeping: The role of cultural and environmental influences on early Neolithic animal exploitation in Europe. *Antiquity, 87*, 1046–1059.

Matthews, B., De Meester, L., Jones, C. G., Ibelings, B. W., Bouma, T. J., Nuutinen, V., et al. (2014). Under niche construction: An operational bridge between ecology, evolution and ecosystem science. *Ecological Monographs, 84*, 245–263.

Mattson, M. P., Allison, D. B., Fontana, L., Harvie, M., Longo, V. D., Malaisse, W. J., Mosley, M., et al. (2014). Meal frequency and timing in health and disease. *Proceedings of the National Academy of Sciences, 111*, 16647–16653.

McMichael, C. H., Piperno, D. R., Bush, M. B., Silman, M. R., Zimmerman, A. R., Raczka, M. F., & Lobato, L. C. (2012). Sparse Pre-Columbian human habitation in western Amazonia. *Science, 336*, 1429–1431.

Modderman, P. J. R. (1988). The linear pottery culture: Diversity in uniformity. *Berichten von het Rijksdienst voor Oudheidkundig Bodemonderzoek, 38*, 63–140.

Nieszery, N. (1995). Linearbandkeramische Gräberfelder in Bayern. In *Internationale Archäologie 16*. Rahden: Leidorf.

O'Brien, M. J., & Bentley, R. A. (2015). The role of food storage in human niche construction: An example from Neolithic Europe. *Environmental Archaeology, 20*, 364–378.

O'Brien, M. J., & Laland, K. N. (2012). Genes, culture, and agriculture: An example of human niche construction. *Current Anthropology, 53*, 434–470.

O'Brien, M. J., Darwent, J., & Lyman, R. L. (2001). Cladistics is useful for reconstructing archaeological phylogenies: Palaeoindian points from the southeastern United States. *Journal of Archaeological Science, 28*, 1115–1136.

Odling-Smee, F. J. (1988). Niche constructing phenotypes. In H. C. Plotkin (Ed.), *The role of behaviour in evolution* (pp. 31–79). Cambridge, MA: MIT Press.

Odling-Smee, F. J., Laland, K. N., & Feldman, M. W. (1996). Niche construction. *American Naturalist, 147*, 641–648.

Odling-Smee, F. J., Laland, K. N., & Feldman, M. W. (2003). Niche construction: The neglected process in evolution. In *Monographs in population biology 37*. Princeton: Princeton University Press.

Peters, J., von den Dreisch, A., & Helmer, D. (2005). The upper Euphrates–Tigris Basin: Cradle of agro-pastoralism? In J. D. Vigne, J. Peters, & D. Helmer (Eds.), *The first steps of animal domestication: New archaeozoological approaches* (pp. 96–124). Oxford: Oxbow.

Pinhasi, R., Fort, J., & Ammerman, A. J. (2005). Tracing the origin and spread of agriculture in Europe. *PLoS Biology, 3*, 2220–2228.

Powell, A., Shennan, S., & Thomas, M. G. (2009). Late Pleistocene demography and the appearance of modern human behavior. *Science, 324*, 1298–1301.

Pringle, H. (2014). The ancient roots of the 1%. *Science, 344*, 822–825.

Richerson, P., Boyd, R., & Henrich, J. (2010). Gene–culture coevolution in the age of genomics. *Proceedings of the National Academy of Sciences, 107*, 8985–8992.

Ruddiman, W. F. (2013). The Anthropocene. *Annual Reviews of Earth and Planetary Sciences, 41*, 4–24.

Salque, M., Bogucki, P., Pyzel, J., Sobkowiak-Tabaka, I., Grygiel, R., Szmyt, M., & Evershed, R. P. (2013). Earliest evidence for cheese making in the sixth millennium BC in northern Europe. *Nature, 493*, 522–525.

Salzman, P. C. (1999). Is inequality universal? *Current Anthropology, 40*, 31–61.

Saqalli, M., Salavert, A., Bréhard, S., Bendrey, R., Vigne, J.-D., & Tresset, A. (2014). Revisiting and modelling the woodland farming system of the early Neolithic Linear Pottery Culture (LBK), 5600–4900 B.C. *Vegetation History and Archaeobotany, 23*, S37–S50.

Scarre, C. (2000). Reply to S. Shennan, "Population, culture history, and the dynamics of culture change". *Current Anthropology, 41*, 827–828.

Scott-Phillips, T. C., Laland, K. N., Shuker, D. M., Dickins, T. E., & West, S. A. (2014). The niche construction perspective: A critical appraisal. *Evolution, 68*, 1231–1243.

Sear, R. (2015). Evolutionary contributions to the study of human fertility. *Population Studies, 69*, S39–S55.

Shennan, S. (2000). Population, culture history, and the dynamics of change. *Current Anthropology, 41*, 811–835.

Shennan, S. (2011a). Descent with modification and the archaeological record. *Philosophical Transactions of the Royal Society B, 366*, 1070–1079.

Shennan, S. (2011b). Property and wealth inequality as cultural niche construction. *Philosophical Transactions of the Royal Society B, 366*, 918–926.

Shennan, S., Downey, S. S., Timpson, A., Edinborough, K., Colledge, S., Kerig, T., Manning, K., & Thomas, M. G. (2013). Regional population collapse followed initial agriculture booms in mid-Holocene Europe. *Nature Communications, 4*, 2486.

Skoglund, P., Malström, H., Raghavan, M., Storå, J., Hall, P., Willerslev, E., Gilbert, M. T. P., et al. (2012). Origins and genetic legacy of Neolithic farmers and hunter–gatherers in Europe. *Science, 336*, 466–469.

Smith, B. D. (2007). Niche construction and the behavioral context of plant and animal domestication. *Evolutionary Anthropology, 16*, 188–199.

Smith, B. D. (2012). A cultural niche construction theory of initial domestication. *Biological Theory, 6*, 260–271.

Smith, B. D. (2016). Neo-Darwinism, niche construction theory, and the initial domestication of plants and animals. *Evolutionary Ecology, 30*, 307–324.

Smith, B. D., & Zeder, M. A. (2013). The onset of the Anthropocene. *Anthropocene, 4*, 8–13.

Stehli, P. (1989). Merzbachtal—Umwelt und Geschichte Einer Bandkermischen Siedlungskammer. *Germania, 67*, 51–76.

Strien, H.-C. (2010). Mobilität in bandkeramischer Zeit im Spiegel der Fernimporte. In D. Gronenborn & J. Petrasch (Eds.), *Die Neolithisierung Mitteleuropas* (pp. 497–508). Mainz: Römisch-Germanisches Zentralmuseum.

Talhelm, T., Zhang, X., Oishi, S., Shimin, C., Duan, D., Lan, X., & Kitayama, S. (2014). Large-scale psychological differences within China explained by rice versus wheat agriculture. *Science, 344*, 603–608.

Tehrani, J. J. (2013). The phylogeny of Little Red Riding Hood. *PLoS One, 8*(11), e78871.

Tehrani, J. J., & Collard, M. (2002). Investigating cultural evolution through biological phylogenetic analyses of Turkmen textiles. *Journal of Anthropological Archaeology, 21*, 443–463.

Timpson, A., Colledge, S., Crema, E., Edinborough, K., Kerig, T., Manning, K., Thomas, M. G., & Shennan, S. (2014). Reconstructing regional demographies of the European Neolithic using 'dates as data'. *Journal of Archaeological Science, 52*, 549–557.

Troy, C. S., MacHugh, D. E., Bailey, J. F., Magee, D. A., Loftus, R. T., Cunningham, P., Chamberlain, A. T., et al. (2001). Genetic evidence for Near-Eastern origins of European cattle. *Nature, 410*, 1088–1091.

Vigne, J.-D., & Helmer, D. (2007). Was milk a 'secondary product' in the Old World Neolithisation process? *Anthropozoologica, 42*, 9–40.

Whitehouse, N. J., Schulting, R. J., McClatchie, M., Barrat, P., McLaughlin, T. R., Bogaard, A., Colledge, S., et al. (2014). Neolithic agriculture on the European western frontier: The boom and bust of early farming in Ireland. *Journal of Archaeological Science, 51*, 181–205.

Williams, G. C. (1992). Gaia, nature worship, and biocentric fallacies. *Quarterly Review of Biology, 67*, 479–486.

Woodbridge, J., Fyfe, R. M., Roberts, N., Downey, S., Edinborough, K., & Shennan, S. (2014). The impact of the Neolithic agricultural transition in Britain. *Journal of Archaeological Science, 51*, 216–224.

Wooding, S. P. (2007). Following the herd. *Nature Genetics, 39*, 7–8.

Zeder, M. A. (2016). Domestication as a model system for niche construction theory. *Evolutionary Ecology, 30*, 325–348.

Zeder, M. A., & Hesse, B. (2000). The initial domestication of goats (*Capra hircus*) in the Zagros Mountains 10,000 years ago. *Science, 287*, 2254–2257.

Zimmermann, A., Hilpert, J., & Wendt, K. P. (2009). Estimations of population density for selected periods between the Neolithic and AD 1800. *Human Biology, 81*, 357–380.

Zvelebil, M. (2006). Mobility, contact, and exchange in the Baltic Sea basin 6000–2000 BC. *Journal of Anthropological Archaeology, 25*, 178–192.

What Can a Multi-agent System Tell Us About the Bantu Expansion 3,000 Years Ago?

Florent Le Néchet, Christophe Coupé, Hélène Mathian, and Lena Sanders

Introduction

The Bantu people (more than 300 million speakers of more than 400 Bantu languages) are the descendants of ancestral populations who, more than 4,000 years ago, started to expand from today's Western Cameroon toward Southern and Eastern Africa (Li et al. 2014), inhabiting today a large part of sub-Saharan Africa. These populations are thought to have brought agriculture and metallurgy with them, with a main episode of expansion occurring around 2,500 years ago. With the accumulation of archeological, linguistic, and genetic evidence, the so-called Bantu expansion is now seen as a complex, fragmented, and long-term phenomenon, made of numerous migrations with their own dynamics, temporality, and spatial range (Grollemund et al. 2015; Vansina 1995).

Many questions remain unresolved today as for the precise routes taken by Bantu farmers. A significant obstacle for them was the rainforest, unsuitable to farming, that they either had to bypass or cross. In doing so, they may have benefited from interactions with the forest foragers they came to meet, and who are today specialists of this environment. How this happened is however still very imprecise, despite accumulating genetic data (Quintana-Murci et al. 2008; Patin et al. 2014).

F. Le Néchet (✉)
Université Paris-Est, Laboratoire Ville Mobilité Transport UMR-T 9403, UPEM, Marne-la-Vallée, France
e-mail: florent.lenechet@u-pem.fr

C. Coupé
Department of Linguistics, The University of Hong Kong, Hong Kong, China

H. Mathian
UMR 5600 Environnement Ville Société, CNRS/Université de Lyon, Lyon, France

L. Sanders
UMR 8504 Géographie-Cités, CNRS-Université Paris I-Université Paris VII, Paris, France

Our work aims to address the previous question with an agent-based model, in order to explore the impact of interactions with forest foragers on the intensity and rhythm of Bantu spatial expansion. Migratory waves of agents representing Bantu groups arrive from the northwestern corner of a stylized spatial grid, move according to their search for resources, and have the opportunity to develop interactions with agents representing forest foragers when they reach forest cells. The development of such a model requires discussions between specialists of the thematic question (linguists, in our case), geographers, and modelers. The aim of this chapter is therefore to detail the main questions which arise when integrating, in an agent-based model and in a simplified way, the experts' hypotheses regarding socio-anthropological factors.

Section "The Bantu Migrations and the Role of Forest Foragers" provides an overview of Bantu farmers' migrations and of their present and past relationship with forest foragers. Section "Simulation Models of Demographic Expansion" advocates for the use of simulation models and summarizes scholarly efforts to model migrations and demographic expansion before section "Overview of the HU.M.E. Model" focuses on our own efforts to first simulate expansion in an empty territory, with the so-called HU.M.E. model. Section "Enriching the HU.M.E. Model to Address the Case of Bantu Migrations" explains how this model was then tweaked to account for the case of Bantu migrations. Section "Discussion: Negotiated Co-construction of the Model" finally offers an epistemological account of the previous modeling attempts.

The Bantu Migrations and the Role of Forest Foragers

Scenarios for the Bantu Migrations

As already said, uncertainties still cloud the past of modern Bantu populations, but various scenarios have been put forward by scholars. Two main scenarios have in particular been opposed on the basis of ethnographic and linguistic data and diverge on the origins of the Bantu populations of Eastern Africa: first, the "east-next-to-the-west" scenario (Bastin et al. 1983, 1999), which argues for two migration routes from the so-called grasslands of today's Cameroon – one going south/southeast and one going first east along the northern border of the equatorial forest before turning southward – and second, the "east-out-of-the-west" scenario (Henrici 1973; Ehret 1973; Heine et al. 1977), which postulates a main migration south/southeast from the grasslands and, later, south of the rainforest, a migratory route going east. Another recent proposal (Bostoen et al. 2015) goes beyond the two previous scenarios. It suggests an early migratory stage, which would have taken place around 4,000 years ago, with limited expansions eastward and southward along the Atlantic coast. These first migrations would not have been associated with agriculture nor metallurgy. It is only 2,500 years ago that larger-scale migrations would have occurred, with farmers able to produce iron tools. According to Grollemund (2012), linguistic evidence leads to reject the "east-next-to-the-west" and "east-out-of-the-west" scenarios in favor of a more complex evolution. Ehret (2001: 40) advocates for "a complex, millenniums-long human history, not of the proto-Bantu diverging at the first stage into ancestral Western and Eastern Bantu peoples, but of successive periods of wider and wider spread of Bantu communities out of the northwest." Furthermore, genetic evidence also seems to discard an "early split" between eastern and western Bantu populations (Plaza et al. 2004; Alves et al. 2011). As explained by Alves et al. (2011: 35–36), "the spreading of Bantu languages is better portrayed as a gradual unfolding of interconnected populations than a series of successive bifurcations involving small sized groups. Alternative models that place the ancestors of subequatorial Bantu peoples at the southern outskirts of the rainforest, shortening divergence times and increasing opportunities for gene flow, seem to better fit the observed genetic patterns."

Among the factors put forward to explain the causes and routes of migration, environmental constraints are carefully scrutinized. Recent data suggest the beginning of a climatic crisis around 4,000 years ago, and a reduction of the area occupied by rainforests in favor of savannahs. A more dramatic change would then have occurred 2,500 years ago, with an amplification of seasonal change and the growth of mixed environments comprising forests of pioneer species of trees (Ngomanda et al. 2009). These historical depths are in line with the previous migratory scenarios. In particular, environmental changes 2,500 years ago would have offered favorable conditions for farming (Bostoen 2006): pioneer forests were easier to cut down during migrations than mature forests and offered space for agriculture, in parallel to a variety of wild plants (Neumann et al. 2012).

How much the forest was an obstacle to early Bantu farmers is a significant question. Unsuitability for farming was one issue, but more generally, they were perhaps as reluctant as today's farmers to enter an environment considered unwelcoming and dangerous. Getting around forest-covered areas was then a reasonable behavior, and climatic change actually opened passages that could serve as shortcuts between the north and south sides of the equator (Schwartz 1992). Linguistic studies of the names of pioneer species suggest that it was indeed the case (Bostoen et al. 2013). Opening 3,000 years ago, the Sangha River interval, separating western and eastern forest massifs (Maley 2001), would have been such a passage, as seen in Fig. 1.

If many groups likely bypassed the forest, others may however have crossed it, although at slower migratory rates (Grollemund et al. 2015). Rivers could have been attractive routes and have influenced the rhythm and patterns of migrations (Russell et al. 2014). Another factor may however have played a greater role: the interactions with the forest foragers who were already occupying Equatorial Africa.

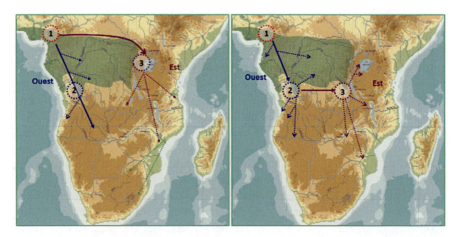

Fig. 1 Possible migratory routes either around or through the rainforest (from Grollemund 2012). Green areas correspond to the patches of rainforest during the main stage of the Bantu expansion, around 2,500 years ago. Arrows correspond to Bantu farmers' plausible migratory paths. Those in the middle relate to the Sangha River interval. Left map: "east-next-to-the-west-model" hypothesis. Right map: "east-out-of-the-west-model" hypothesis. These routes are diachronic reconstructions from synchronic (contemporary) linguistic data

Present and Past Relationships to Forest Foragers

Today, Bantu farmers live close to, and interact with, forest foragers in several regions of Equatorial Africa. These forest foragers (FF thereafter), often called Pygmies, number in the few hundreds of thousands and can be grouped in around 20 groups which display a variety of lifestyles, living conditions, languages, and relationships with farmers (Bahuchet 1991, 2012). Assimilation to the latter can be stronger or weaker, but FF most often share deep social and economic bonds with them. They have done so in the past, to the extent that there is no linguistic evidence today of ancestral hunter-gatherer languages. Beyond their small stature, FF share a cultural trait, namely, their position as specialists of the forest, which constitutes the source of part of their subsistence but also the center of their spiritual life (Ichikawa 2004).

FF partly live from hunting and gathering in the forest but could not survive in isolation into it. They get agricultural products but also iron tools and potteries from the Bantu, in exchange from forest resources. Manual labor for the farmers is also part of the equation. A symbiotic relationship has very likely been in place for a long time, although it has degraded significantly from the onset of the colonial era, with a growing asymmetry between farmers and FF in favor of the former. Today, strong negative social representations from farmers and poor living conditions are widespread among FF (Bahuchet and Guillaume 1982), but were not necessarily present at first. Recent genetic studies show an increase in gene exchanges between Bantu farmers and FF during the last 1,000 years (Patin et al. 2014). This raises the question of the evolution of the interaction between these two groups. Were FF living on the outskirts of the forest prior to meeting farmers, extracting their subsistence both from the forest and the savannah – which makes sense if living solely from the forest is impossible? Did they then reinforce their expertise of, and relationship to, the forest as a result of the contact, in order to establish complementary and reinforcing skills and knowledge between the two groups? Or was the contact much less of a transforming event? Were the relationships then much more egalitarian than they are today, as it is suggested by the farmers' traditional stories (Bahuchet and Guillaume 1982)?

The nature of the initial contacts between Bantu and FF is difficult to assess and depends on the answers to the previous questions. In any case, FF may have played a significant role in the migratory routes taken by groups of farmers. By helping these farmers to get resource from the forest and guiding them through an environment they were much at ease with, they may have favored direct routes through forest massifs. Simultaneously, one can imagine that strengthening interactions with FF, more inclined to local mobility than large-scale expansions driven by the need of new lands, could have in some cases slowed down the migratory processes.

Different scenarios for the migrations and interactions with FF are therefore available, with their respective articulations of causes and consequences. The questions are raised of how to assess their relevance and whether one can make a reasonable choice among them. The modeling framework described in the next sections is an attempt to answer these questions.

Simulation Models of Demographic Expansion

In the absence of consensual evidence, how to make progress in relating putative past events involving early Bantu migrating farmers to the current visible output of migrations, i.e. an expansion including most of sub-Saharan Africa? In fact, results discussed in section "The Bantu Migrations and the Role of Forest Foragers" are principally based on analyses of genetic and linguistic data. On the other hand, archeological data are too fragmentary and incomplete to answer the raised questions. Other ways than data analyses have then to be explored to find answers. Lake (2014) distinguishes models based on statistical analyses of data and models using simulation to develop "thought experiments" in a more exploratory perspective. These two families of models, corresponding to different epistemologies, are complementary: the former are useful for testing hypotheses when appropriate data exist, and the latter may support the initial construction of theories. In this latter case, the most important is not to reproduce empirical facts exactly, but to reflect on the processes which may underlie these facts. Premo (2006), for example, underlines that agent-based models are tools for exploring "alternative cultural histories" based on "what-if" scenarios. He proposes to replace the classical question "What happened in region X during period Y?" with "How likely is it that behavior Q or trait Z would evolve in the population in region X during period Y given a wide range of plausible environmental conditions and alternative histories?". While the aim of a statistical model is to find out the relations that did exist between phenomena, the simulation model can be used as a laboratory for exploring the long-term effects of different forms of relations.

The previous alternative methodology could therefore help bridging the gap between local mechanisms and behaviors and resulting patterns at a larger scale. It is argued below, through a series of examples attempting to model demographic expansion in different contexts, that such kind of modeling approach can indeed be a meaningful answer. Section "Simulation Models of Demographic Expansion" summarizes these attempts, which provide the background of our own attempt, described in sections "Overview of the HU.M.E. Model" and "Enriching the HU.M.E. Model to Address the Case of Bantu Migrations". A first group of models concerns peopling of new land; a second group concentrates on the role of interactions in such a process.

Modeling the Peopling of Empty Land

The peopling of new land is a classic case of long-distance migration in prehistoric periods: human groups, searching for resources, enter and settle in a previously unoccupied space. Many models have focused on the diffusion mechanism associated to such a colonization process (Hazelwood and Steele 2004). One of the most generic is the model developed by Young (2002). This model is agent-based and formalized at the level of individuals. A number of agents are entering an empty, homogeneous, and isotropic spatial grid. The peopling process is interpreted as the result of individuals' actions, and the model is based on rules associated to four processes: (i) two demographic processes, (ii) a process of competition, and (iii) a process of migration. The four processes are applied to each agent at each iteration, and when a move occurs, it is in a random direction. The author shows that the combination of a high growth rate and a low migration rate results in a wave front which form is reminiscent of the propagation of Neolithic farming. On the opposite, a slow growth rate and a high probability of mobility lead to a diffuse and sparse peopling of the whole space, like the colonization of Australia during the Pleistocene. The choice made by Young (2002) of a very stylized space contributes to the genericity of the model. It is thus adapted to reflect on the main mechanisms of diffusion operating at large spatiotemporal scales (demography and propensity to move), but not to explore hypotheses about the effects of spatial heterogeneity in terms of resources and topography. Because it deliberately avoids to relate the decision and direction of movement to any geographical characteristic of space or any specific cause, it cannot efficiently be used in a dialogue with thematic experts.

Indeed, the reasons to move could be of different kinds: a response to disease or violence (Kohler et al. 2014), demographic overcrowding, the search for new resources (Parisi et al. 2008), etc. In this latter case, resource is explicitly formalized in the model, and its scarcity is the main driver for mobility. This approach implies to take into account the environmental features of the modeled space in order to estimate the carrying capacity of different locations. The same two mechanisms as used in the Young model (growth and migration) are combined, but in this case are imbedded in a heterogeneous space. In fact, continuous or discrete approaches can be used to formalize such combination of mechanisms.

Continuous models include the well-known Fisher-Skellam model (reaction-diffusion) in the field of population ecology. It is based on a deterministic differential equation expressing the evolution of population density in space according to two terms: a term relating to the local demographic growth according to a carrying capacity (using the mathematical logistic function which suits well a growth phenomenon that saturates toward a maximum), and a second term connected to the spatial diffusion of the population. Ammerman and Cavalli-Sforza (1984) used this approach to model the spread of Neolithic farmers, as a front wave, in Europe between 8,000 and 4,000 BCE. Davison et al. (2006), applying a similar approach, considered altitude, latitude, and sea travel to estimate carrying capacities and mobility, highlighting the role of waterways in the spread of agriculture in Europe. Hazelwood and Steele (2004) finally took the same approach to reproduce travelling waves of different shapes and velocities in the first peopling of the Americas. They interpreted different outputs of their model as signatures of types of population expansion, which they then compared to archeological records.

As previously said, the previous mechanisms can also be formalized using a discrete approach. In Parisi et al. (2008), for example, space is represented by a grid, and people move when resources available on the cell where they are located are insufficient with respect to their needs. The model rests on cellular automata (CA), each cell being characterized by a carrying capacity, estimated from its farming potential, and a quantity of population. The initial situation is assumed to occur around 9,000 years ago, with a single cell (corresponding to South-Western Anatolia) occupied by a population of farmers. The main rule of the model is based on the triggering effect of the lack of resources: when the needs of the population exceed available resources, part of it moves to a neighboring unoccupied cell, randomly chosen among those with farming potential. The diffusion process thus gradually drives the populations away from their starting point with a specific spatial pattern.

During its migrations out of Africa and Eurasia, i.e., to Australia, the Americas, and later Oceania, our species colonized empty lands (there were other human species in Africa, Europe, and Asia, and our interactions with them are heavily debated), but empty space became the exception rather than the rule as the peopling of the whole planet made progress. This sets limits to the previous models and highlights the need to further account for interactions between human groups in occupied spaces.

Modeling the Peopling of Already Inhabited Land: The Crucial Role of Interactions

In the models described above, the driving evolutionary force is the interaction between people and their environment. Aside from these interactions linked to resources, interactions between humans may also have an impact on the shape and speed of migrations. Two completely different kinds of interactions are generally formalized in models referring to peopling process: (i) exchanges of resources or techniques between groups and (ii) exchanges of genes. In the first case, the modeler is often interested in diffusion processes, and a time granularity corresponding to the year is then well adapted to capture the rhythm of propagation (Ortega et al. 2016). In the second case, the question at hand most often refers to the long-term consequences of genetic mixing, and the generation is then a more suitable time step (Barton and Riel-Salvatore 2012; Currat and Excoffier 2005; Rasteiro et al. 2012). However, the duration of the studied phenomena is not the only criterion of choice for time units. The Neolithization process in Europe, for example, has been modeled using different time steps and different levels of explicitation of the interactions between immigrating flows of farmers and native forest foragers. In some models, these interactions are simply left aside. It is, for example, the case in Parisi et al. (2008)'s model, which focuses solely on the consequences of resource exploitation on farmers's expansion, which is simulated at the time scale of the year. In other models, the effect of interactions is formalized through change of status, with a proportion of forest foragers becoming farmers at each time step (cf. Ammerman and Cavalli-Sforza 1984). Yet in others, the interactions explicitly point at the genetic mixing of agents of different kinds, simulated at the time scale of a generation. Currat and Excoffier (2005) used this latter approach to assess the two main scenarios for the Neolithization process in Europe: (i) the demic diffusion hypothesis, with farmers from the Middle East migrating to Europe – a scenario which advocates for genetic mixing between migrating and autochthone populations, and (ii) the cultural diffusion hypothesis, which assumes transmission and spreading of farming techniques with much less movement of populations.

A similar approach was used by Barton and Riel-Salvatore (2012) and Barton et al. (2011) when they implemented a multi-agent system (MAS) in order to study the coexistence of Neanderthal and *Homo sapiens* populations during the Upper Pleistocene. It is well-known that this coexistence led in the long run to the disappearance of the former, around 35,000 years ago. The model was used as a laboratory to explore the long-term demographic consequences of different hypothetical biological differences between the two species, and of different types of interactions between their members. More precisely, the authors evaluated the outcome of different physical aptitudes between the species, of various behaviors in terms of assortative mating, and of the different extensions of their resource catching areas (home range). In their multi-agent system, Neanderthal agents and *Homo sapiens* agents were, respectively, distributed in the western and eastern part of a spatial grid representing Europe. Each agent was characterized by ten allele pairs. Agents searched for resources in their home range. If one of them met another agent, the couple could give birth, following a random drawing, to a new agent inheriting a combination of its parents' genomes. New agents were put in an unoccupied cell, near but outside their parents' home range. Agents did not migrate, and the spatial diffusion of the population was solely due to the birth and relocation of offspring. The model ran for 1500 iterations, each corresponding to one generation, with a total time for the simulation therefore corresponding to around 30,000–35,000 years. The simulations showed that the disappearance of Neanderthal could be due either to a difference of physical fitness or, more interestingly, to a higher level of interbreeding due to the extension of the agents' catching areas (as a response to climatic cooling). The long-term consequence of such process is an "extinction through hybridization."

These examples show various options that have been investigated with simulation models and that can serve as an anchor for our attempt to better understand the case of Bantu migrations and the role of contacts with forest foragers. The simulation models based on MAS seem particularly appropriate. A MAS relies on agents, which relate intuitively well to the groups of farmers and forest foragers. The behaviors of agents, and especially their interactions with their environment and with other agents, are defined by rules which are local but can lead to emergent collective behaviors through self-organizing processes. In the case of Bantu farmers, groups had no idea of what was going on except in their neighborhood – there were no means of distant communication, nor satellite imagery. There was very likely no centralized coordination of the whole Bantu expansion, which global shape and rhythm therefore derived in an emergent way from the myriads of local migratory behaviors and interactions. There is thus a "natural" match between MAS and the situation which took place in Central Africa a few thousands years ago. MAS, as simulation models in general, also offer the possibility to deal with the intrinsic stochasticity of the processes, by repetitions of simulations with the same initial conditions and rules. They thus provide a testbed for various hypotheses – Premo's "what-if" scenarios – regarding interactions between Bantu farmers and FF and their consequences on the intensity and rhythm of the Bantu expansion southward (Premo 2006; Barton and Riel-Salvatore 2012). It is important to state that the model must rest as much as possible on well-established thematic knowledge, although simplification and choices are unavoidable.

Sections "Overview of the HU.M.E. Model" and "Enriching the HU.M.E. Model to Address the Case of Bantu Migrations" illustrate the approach we took, which consisted in breaking down the complexity of the situation by first studying migratory patterns in an empty but heterogeneous space, focusing on the interaction with the environment (section "Overview of the HU.M.E. Model"), then adding human interactions, and focusing on their possible further impact (section " Enriching the HU.M.E. Model to Address the Case of Bantu Migrations").

Overview of the HU.M.E. Model

As previously said, Bantu migration routes are not known with certainty: did farmers move along corridors of savannah? Along the sea or rivers? Did they cross or get around the forest massifs? Data are lacking for these remote historical periods. In such an epistemological context, a MAS can serve as a laboratory to explore different hypotheses on the Bantu progression in a diversified space (to keep things simple, made of savannahs and forests) and on the role of interactions with FF on the likelihood and rhythm of such migrations. As seen previously, these interactions could cover various aspects, and our choice has been to focus on the exchange of food resources. This is particularly meaningful in light of the limited skills Bantu farmers have to survive in the forest. The aim is then to explore the consequence of getting food from FF on the ability of Bantu groups to cross the forest and reach savannah beyond it. Migrants were of course unaware of the existence of these savannah areas before reaching them, and were not planning their moves in order to reach them. It is plausible that these moves were primarily guided by a local search for resources, in which FF could play a significant role. Reaching the areas south of the rain forest can thus be interpreted as an "emergence" rather than the consequence of conscious and intended actions. This reasonable hypothesis meets the philosophy of MAS based on local rules and a bottom-up principle, without overarching control.

The acronym HU.M.E. stands for human migration and environment. As described below, the HU.M.E. model articulates mechanisms underlying the peopling of an unoccupied land by human groups, in a highly stylized fashion and with an intermediate level of abstraction according to the KISS/KIDS classification of

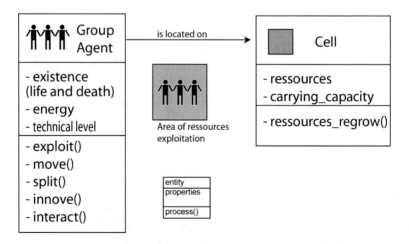

Fig. 2 Entities, properties, relations, and processes formalized in the HU.M.E. model

MAS models (Edmonds and Moss 2004): it incorporates some rules that directly come from the exchange with thematic experts regarding the Bantu expansion and yet remains quite theoretical, generic, and exploratory.

The HU.M.E. model has been developed following a progressive method. First, the different pieces necessary to model the peopling of a new land are brought together. The model focuses on agents searching for resources and moving in an evolving environment. It uses a MAS to describe a process of peopling of unoccupied land, starting with the arrival of migratory groups coming from outside of the target space. The aim of this first simple model is to highlight the effects of simple rules, describing the interactions between groups and the dynamics of the exploitation of resources, on the speed and the spatial configuration of the peopling. Space is formalized discretely through a grid (52 cells × 52 cells) where each cell is characterized by a level of resources, referring to a quantity of biomass. The size of the cells corresponds to the catchment area which can reasonably be covered on foot by hunters or gatherers (around 15 km × 15 km). The resource regenerates progressively after having been harvested. Human groups are represented in the model by "agents" following the approach framework that we call "group-agent" (cf. Fig. 2). They are characterized by two properties – a technical level and an energy level – and their behavior is formalized through five mechanisms:

- Harvesting resources: each group-agent has the need and ability to harvest the resources of its cell of location according to its technical level. As long as a group is able to stay on a cell, it accumulates energy which will be used when it moves.
- Moving from one cell to another: when resources are missing on the cell of location, the group-agent moves to a neighboring cell. If resources are missing there, it will move again at the next time step if it still has "energy" enough. The direction of the move is random, and there is no anticipation according to the cells' resource level. When resource is missing and there is no more energy, the group-agent dies.
- Splitting up: when the energy level is high – a sign of strength – the group may split up in two and give rise to two new groups. This dividing mechanism stands for the demographic growth process.
- Innovating: the group-agents have the ability to innovate, which means to improve their technical level and therefore their ability to extract the resources of the cell where they are located. The harvest is nevertheless limited by the carrying capacity of the cell (cf. Fig. 2).
- Interacting with other groups: when two groups are located on the same cell, two kinds of interactions are possible: (i) an indirect one through competition for resources and (ii) a direct one, during which the group with the lower technical level may acquires the technical level of the other group by imitation.

The behavioral rules connected to movement and to demographic growth are of course designed to express the growth of the Bantu populations which occurred during their expansion. In a similar fashion, the notions of innovation and transfer of technical knowledge reflect the likely gradual development of a variety of techniques in groups of Bantu farmers, whether related to agriculture, metallurgy, fishing, etc. Finally, despite lacking a global vision of their environment, one can say that groups understand its heterogeneous structure in that they accumulate resources in prevision of future needs during migration, although they do not choose the best new locations when they move.

This simple model is sufficient to reproduce the population spread associated to a peopling process such as the Bantu expansion. In the initial state, space is homogeneous, all cells having the same level of biomass. This level will decrease progressively as a cell is occupied and harvested by groups. It will also be gradually replenished once the cell is abandoned by these groups. Migrating groups arrive in successive waves from the northwest corner of the grid during the first steps of the simulation.

The model served as a base of discussion between modelers and experts of the thematic domain. The aim was to reflect on the thematic signification of the introduced mechanisms, in order to identify possible oversimplifications and to validate the global modeling framework. Two examples illustrate this discussion:

- "Moving without reason": resource scarcity is not the only reason bringing human groups to move – intra- or intergroup conflicts or cultural beliefs may play a role in the decision process. A model based solely on constraints of resource was considered to be too deterministic. A "probability to move without reason" was thus introduced. Even if this probability was low, one could expect consequences in the long run through changing potentials of interaction and path dependency. The terminology "moving without reason" introduced by the modelers was actually rather aberrant for the thematic experts, who instead argued for the multiplicity of plausible reasons. Indeed, following a principle of parsimony, the modelers accounted for these different reasons with a simple random factor. "Moving without reason" thus means in fact "moving for a reason other than resource scarcity." This anecdotal example illustrates everyday possible misunderstanding when thematic experts and modelers co-construct a model.

- "Energy": the concept of "energy" is used in a metaphoric way to represent the hypothesis that a group staying for a time at the same location can rest and benefit from a time of preparation that facilitates the next moves. Therefore it mainly concerns the farmers as hunters-gatherers' way of life is not based on storage of possible surplus. Working on animal populations, Nonaka and Holme (2007) have highlighted the relevance of a MAS to model agents' moves in search of resources in a heterogeneous landscape. Energy plays a key role in their model. From a conceptual point of view, the situation is the same in the HU.M.E. model: indeed, resources become heterogeneous due to exploitation and regeneration, and predator-agents move and consume. It thus seemed interesting to use a similar notion of energy, and derive some mechanisms of the model from it.

Enriching the HU.M.E. Model to Address the Case of Bantu Migrations

In the HU.M.E. model described above, the land where the migrants arrive is unoccupied, whereas in the case of the Bantu expansion, farmers actually encountered groups of forest foragers (FF). In order to explore the effects of their interactions, the model needs to be enriched. Three successive enrichments are presented in this section: (i) first, two distinct types of groups, Bantu farmers and FF, are introduced; in parallel, the environment is split into two distinct types of land, forest and savannah; (ii) second, the way each type of groups interacts with the two distinct types of environment is elaborated; and (iii) last, rules of interaction between Bantu agents and FF agents are specified.

Introducing Diverse Populations and Environments

The thematic experts clearly stated the heterogeneity of human groups in terms of relation to the environment and movement. In the model, we therefore created two types of agents: Bantu farmers and FF. Both types share the same general attributes, but some parameters defining their behaviors have been differentiated upon discussion with the thematic experts: the propensity to innovate is stronger in Bantu groups than in FF groups; Bantu groups store "energy" but FF groups don't; and demographic expansion is set at a very low rate for FF (stable population, as commonly found for forest foragers) and at a high rate for Bantu (taking into account the demographic expansion of Bantu groups around 2,500 years ago).

In the model, the grid has been adapted to include two types of environment: savannah and forest. To achieve this, we used "biomass" of cells that expresses the amount of resources available for human groups. Next section details how group-agents can exploit biomass. Note that the modeler can design different initial configurations of the grid, depending upon the spatial distribution of biomass between cells. For instance, zones representing migratory obstacles can be created by initializing cells with very low or zero biomass.

Differentiating Resource Exploitation for Bantu and FF Groups

In addition to the storage of food previously described in the HU.M.E. model, we introduced at the group-agent level a capacity of anticipation with respect to the heterogeneity of space: before moving, group-agents assess the resources available in neighboring cells and are more likely to move to cells which better accommodate their needs in terms of resource consumption. More precisely, we used the biomass attribute of the cells to compute a score, based on a utility function, for each neighboring cell.

The utility of a cell D for a group with technical level T is noted $U_T(D)$. It is calculated via a function f, the potential of exploitation, which depends upon the technical level T of the group and the biomass b_D of the cell D. The mathematical form of the f function is detailed below. It is important to note that after discussing with the thematic experts, we chose in our model to define higher values of biomass for forest cells than for savannah cells:

$$U_T(D) = f(T, b_D)$$

We actually created two different f functions for Bantu groups and FF groups, in order to express their different abilities to exploit resources of forest and savannah environments. Denoting C the "culture" of a group-agent, i.e., Bantu or FF, we reformulated the utility function as follows:

$$U_{C,T}(D) = f_C(T, b_D)$$

There are two major differences in the calculation of the utility function between Bantu and FF groups. First, as described previously, the technical level of FF groups is constant over time, while it can change over time for Bantu groups. Secondly, the ability to exploit resources is different for Bantu groups and FF groups. For the former, the most favorable environments are those with intermediate levels of biomass, which correspond to savannahs and not to forests, since thematic experts stated that Bantu farmers have limited skills to gather resources in forest environments. We therefore implemented a nonlinear utility function for Bantu groups: f_B. f_B increases linearly with biomass b up to a threshold b_f and then decreases drastically. Low values of b_f then imply a strong repulsive effect of forest environments. For FF groups, f_{FF} linearly increases without any threshold, which indicates that FF groups value higher levels of biomass and thus especially forest environments (Fig. 3).

Introducing Specific Interactions Between Bantu and FF Groups

Next, we formalized the direct interactions between Bantu groups and FF groups. In the initial model (section "Overview of the HU.M.E. Model"), interactions between groups only occurred when two group-agents were located in the same cell. On the one hand, because of the competition to extract the same resources, one of the groups (chosen randomly) had to migrate when resources became insufficient for the two groups. On the other hand, proximity could lead to a technological transfer from the more technologically advanced group to the less advanced one. This type

What Can a Multi-agent System Tell Us About the Bantu Expansion 3,000 Years Ago?

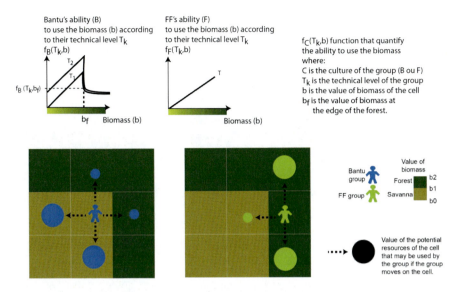

Fig. 3 Differentiated abilities to extract resource as a function of culture and of the type of cell (forest or savannah)

of interaction was maintained in the enriched model, and a rule of direct interaction was added between Bantu groups and FF groups.

According to thematic experts, Bantu and FF have dramatically different knowledge and abilities, and a strong hypothesis from the literature is that these differences fostered situations of "mutual dependence" between the two types of group. As previously said, FF groups hardly live today in autarky in the forest and exchange forest products and labor work against a variety of food and tools with Bantu farmers. Bahuchet (1991) has suggested that this relationship has deep roots in history and is much more asymmetrical today than it was in the past. It remains however unknown how symbiotic the relationship was during the first contacts between Bantu and FF.

In the model, we formalized the previous relationship between Bantu farmers and FF by introducing a possible link between group-agents of the two different types. This link appeared or disappeared under specific conditions and was considered as "elastic." This aimed to account for the reciprocity of exchanges between Bantu and FF groups, and a Bantu group could be attracted by a FF group and vice versa.

We formalized mostly one aspect of this reciprocal relationship, namely, what was related to the exploitation of resources, and left aside much of the likely complexity of the past interactions. More precisely, in the case where groups of different types were located on adjacent cells and were not already involved in another "cross-type of groups" interaction, a link was created between them (i.e., each group could only be linked to a single other group). Each group then included in its evaluation of neighboring cells what its relationship with the other group brought in terms of supplementary resources.

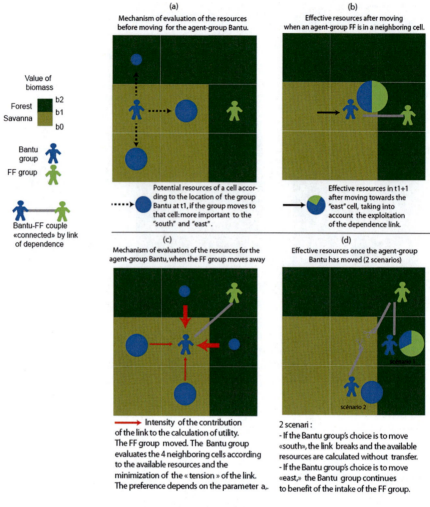

Fig. 4 Possible direct relationship between neighboring groups of Bantu farmers and forest foragers

As depicted in Fig. 4, a connection between groups had two consequences:

- First, Bantu groups obtained additional resources from FF, the amount being defined by a parameter Ω. All other things being equal, this led them to stay longer in the same cell. The resources brought by FF were not removed from their own; this choice was justified by the thematic experts by the idea that these resources were easily extracted from the forest by FF, without depleting their own stocks – i.e., a small cost of extraction for high-value resources.
- Second, in terms of rules of movement, the elastic link played a role when choosing the next cell to which to migrate. Indeed, when one of the two group-agents

moved, an attractive force toward the other made the agent try to minimize the future distance with the "attached" group. However, over a certain distance, the link broke and the two groups regained independence.

The f_B and f_{FF} functions were modified to reflect the previous consequences. The extra resources available to Bantu groups when in contact with FF were included in the calculation of the potential resources at a given cell. Also, the propensity to minimize the distance between connected groups was formalized through a function $g(L) = 1/(L + 1)$, where L was the length of the link. The f_C functions were formalized via a Cobb-Douglas function that weights the relative importance of maximizing resources and minimizing distance with a parameter a. The utility functions then became:

$$U_{C,T}(D) = f_C(T, b_D)^a g(L)^{1-a}$$

Direct interactions were obviously meaningful with respect to the extent to which Bantu migratory routes were influenced by relationships with FF groups: did they lead to faster or slower rates of migration? Did they determine the success of episodes of expansion toward the south and the east?

One can say that despite the reciprocal nature of the relationship introduced in the model, the transfer of resources from FF to Bantu, and not vice versa, manifests an asymmetry in the interactions. We will come back to this choice of mixing symmetrical and asymmetrical aspects of the relationship in the discussion.

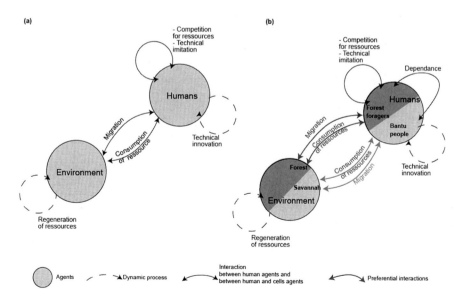

Fig. 5 Schematic representation of the mechanisms and interactions between the entities of the (**a**) initial model (left) and (**b**) enriched model (right)

Figure 5 summarizes the difference between the initial and enriched HU.M.E. models and how the former was modified to create the latter. On the left, the main features of the initial model are represented, focusing on the dynamics of interactions of the FF groups with the environment and with other groups. On the right, the enriched model includes two different types of cells and the existence of two different kinds of groups, Bantu and FF, each with its own strategy to harvest resources.

Designing an Experimental Protocol

The purpose of this section is to articulate the content of the previous sections with an experimental protocol allowing to test our research hypotheses, in particular that of the influence of FF groups on the rhythm and structure of Bantu migrations.

In our opinion, the design of such an experimental protocol is one of the key elements that must be negotiated between modelers and thematic experts.

Figure 6 summarizes our ideas to this end. It shows our reference initial configuration, which we designed as a highly stylized representation of central Africa, extending north and south of the equatorial rainforest. It is an environment mainly made of savannahs, split by a forest zone which plays the role of an obstacle along Bantu migration routes originating in the northwest. A corridor of savannah has been created to account for the Sangha River interval, a simplification discussed in section "Discussion: Negotiated Co-construction of the Model".

A simulation starts at *t0* with a Bantu population, initially of low demographic importance, located in the northwestern part of the territory. Bantu groups then spread throughout the territory as resources decrease in their initial locations due to their exploitation (*t1*). The objective of the experiment is to identify the various

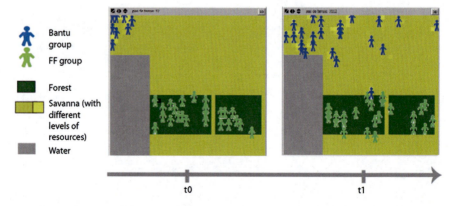

Fig. 6 Initial and later steps of the simulation of the enriched HU.M.E. model. At *t0*, only two values of biomass can be found in the grid: a low value for savannah cells (b_s) and a high value for forest cells (b_f). As resources are extracted and regenerate, cells diverge and a continuum of values emerges

possible paths (bypassing the forest by the east, crossing through the forest, taking the central savannah corridor, etc.) and the way the presence of FF groups influences the speed or paths of the Bantu expansion.

Simulation results (Coupé et al. 2017) show that interactions with FF group-agents tend to favor attempts by Bantu groups to cut directly through the forest and generally tend to slow down the speed of this crossing. The introduction of specific interaction rules between Bantu and FF group-agents, with a "mutual attraction effect," thus makes it possible to cross the forest, although slowly, which is in agreement with Grollemund's proposals (Grollemund 2012) (see section "The Bantu Migrations and the Role of Forest Foragers").

The rules of our model are focused on local migrations, step by step, in one direction or another given the surrounding environment. No "intention" exists of crossing the equatorial forest southward to occupy the savanna territories beyond it. There is in fact no knowledge of the very existence of these savannahs by the Bantu group-agents. Their crossing of the forest and the southern expansion are rather emerging properties of the model. This is why the results of our simulation support the idea that the crossing of the forest is possible without planification by the Bantu groups. According to the exploratory results of the simulation model, the presence of FF group-agents in the forest has two consequences. First, it increases the likelihood of (unintentional) attempts to cross the forest. Hence, at the Bantu group-agent level, the likelihood to choose a forest cell as the next destination is higher with the extra resources potentially provided by a connected FF group-agent. Second, the presence of FF group-agents tends to increase the time necessary to achieve the crossing – at the Bantu group-agent level, the likelihood to stay in a given cell is increased due to the additional resources provided by the close-by FF group-agent. However, according to the simulation results, the presence of FF groups has no effect on the likelihood of Bantu groups to succeed in their attempts to cross the forest – at the Bantu group-agent level, it neither gives an impetus to store more energy nor reduces the likelihood to die from a lack of resources. Overall, as more Bantu group-agents will enter the forest, more will reach the southern region, leading to a more intense Bantu expansion when taking into account the interaction with FF group-agents.

Discussion: Negotiated Co-construction of the Model

The first formalization of the HU.M.E. model refers to a very generic process associated to the exit out of Africa and is as such based on a certain level of abstraction. Calibrated given the speed of movement of the groups, it has first been used without specifying neither temporal and spatial scales, nor specific geographic locations (Coupé et al. 2017).

The second, "enriched", version of the model is a specification of HU.M.E. to consider a given population (Bantu farmers) in a given geographical area (sub-Saharan

Africa). In this model, we included two types of groups with different behaviors toward environmental resources, which created the need to formalize interactions that were both specific to our case study – to account for some specific behaviors – and had some genericity – the rules we introduced in the model had to make sense over a period of approximately 3,000 years. This specification was co-constructed: modelers and thematic experts sought together to identify the salient elements to be integrated into the model, avoiding at the same time the risks of oversimplification and over-specification.

Thus, we were able to obtain a stylized and thematically meaningful model. While the basic rules of the enriched model correspond to a consensus, as a result of the co-construction process, we faced the challenge to conciliate various points of view during this process. These points of view connected to various disciplinary backgrounds and were mainly related to the way space and time were implicitly understood in the model. It was less a difference between thematic experts and modelers' points of view than a difference between linguists and geographers' points of view.

In our research, the object of the modeling is clearly delimited by a spatial scale (the space of the Bantu expansion shown in Fig. 1) and a temporal scale (4,000 years, i.e., the approximate duration of this expansion). As the spatiotemporal context was easily shared by all disciplines, one would have expected the formalization of space and time to be straightforward. However, the spatial and temporal granularities associated with thematical knowledge, their synchronization, interlocking, and interrelations, were one of the main sources of misunderstanding during the co-construction of the model. An example can illustrate dual interpretations due to different disciplinary habits between geographers and linguists. The following quotation from section "The Bantu Migrations and the Role of Forest Foragers" describes the empirical knowledge at the root of this work. It concerns the plausible routes followed by Bantu groups during their expansion southward:

> The spreading of Bantu languages is better portrayed as a gradual unfolding of interconnected populations than a series of successive bifurcations involving small sized groups. (Alves et al. 2011: 35–36)

This sentence was quoted by a linguist whose work is largely influenced by the genetic and linguistic literature. The term "interconnected" was used because empirical observations show traces of genetic mixing between different Bantu groups involved in the southward expansion. It remains however uncertain whether this mixing took place over a long time period (e.g., hundreds of generations) or more intensively over a shorter time period. However, the geographers were not initially aware of this uncertainty. With their own habits, they rather interpreted the same term "interconnected" as suggesting the existence of information transmitted by individuals from one group to another. This gap in the interpretation of the term "interconnected" may have influenced the formalization of the model: in the enriched model, some variables, such as the technical level which takes place at group level, can spread through interactions between groups.

Ambiguity also lies in the metaphoric expression of "routes followed by groups," which is, in the field of geography, implicitly associated with notions such as changes of location, which could be traced with historical data. The knowledge of linguists is of a different nature, since it is based on linguistic and genetic data that are localized and coming from contemporary analyses. Linguists are able to construct phylogenetic proximity graphs – proximity between the languages spoken today – that are then indirectly projected on a geographical space and allow for diachronic reconstructions. In doing so, linguists generate knowledge through inductive techniques and summarize this knowledge in various forms, including maps. In their phylo-geographic reconstructions in particular, arrows only inform the reader of diachronic hypotheses which can easily be overinterpreted by geographers who usually associate arrows with spatial traces. This may have influenced the design of the migratory mechanisms in the enriched model. Thus, the notion of "path" or "route" was essential to the experimental setup detailed in section "Designing an Experimental Protocol".

Overall, for the co-construction of the enriched model, the calibration of the spatial quantities has been consensual because of well-established knowledge provided by the thematic experts. For instance, the size of "catchment areas" has been investigated by ethnologists and prehistorians (e.g., Biraben et al. 1997; Hassan 1981), and reasonable suggestions could be made for the model. The calibration of temporal quantities has been more complex: the time step of the model has been set to a year, even though the thematic experts only provided information over much larger time periods – often in the order of a few centuries. It required an interdisciplinary reflection on the meaning of the behavior rules: for instance, the thematic experts could answer questions such as "Is it likely that a given group changes cell approximately once by generation?", thus providing the modelers with validation at a mesoscopic temporal scale. Another source of complexity regarding the calibration of time quantities is the very method from which order of magnitude is grasped by thematic experts, because they mostly use data concerning nowadays population or ethnographic data that have a bit more temporal depth. At the opposite, the model runs over periods of time that we assume to be in the order of thousands of years in the past, at a time where no ethnographic data or direct observation can be used. This is however intrinsically connected to the goal of the model, which is to investigate periods of the past for which little or no ethnographic or archeological knowledge is available.

Additionally, since much of the linguists' knowledge regarding past events comes from the work of anthropologists and ethno-ecologists on contemporary or recent periods, the formalization of rules in the model was influenced by the evolution of the system toward the contemporary situation. Forest-related differences between Bantu and FF are an example: the choice was made in the model to consider, throughout the whole 3,000 years covered by the simulations, the contemporary observations of Bantu farmers' reluctance to get into the forest, and of FF's skills in this environment, although this may have been different in a distant past.

In addition, the initial HU.M.E. model has been calibrated (in terms of cell size and of various parameters associated with the likelihood of migrating) so as to obtain a speed of expansion consistent with empirical estimations (approximately 1 km per year). The orders of magnitude of the parameters begetting the average speed of expansion were maintained in the enriched model, which allowed us to explore the effects of different contexts, including the presence of a "barrier" (the forest) and interactions with FF group-agents, on the progress of migration. The time needed to cross the forest was thus considered as an emergent property in this second model, whereas in the first model diffusion time was actually used in the calibration process.

One should note that interpreting time is difficult in such a microscopic model, and probably constitutes one of the biases of this type of approach: because the time of the simulation is a linear, mechanical time, comparable to a counter, it should not be interpreted absolutely, both in dates and duration, but in relative terms. In reality, indeed, groups do not gather each year at a precise date to decide whether to migrate or not and in which direction. However, this simplification seemed acceptable because of the very high number of iterations (a few thousands) in a simulation, and because of the stochasticity introduced in the model. Overall, we can give relative interpretations of the migration rate between simulations, but not interpretations in absolute length of the periods elapsed.

From a thematic point of view, the environment is heterogeneous. Our formalization into two types of land, "savannah" and "forest," and therefore two types of land use, resulted from a consensus between modelers and thematic experts. This differentiation is a cornerstone of the model and is required to be able to implement distinct behaviors in Bantu group-agents and FF group-agents – once again, the forest was considered as a hostile environment for the Bantu group-agents, whereas it represented a favorable area for the FF group-agents. Also, the savannah/forest dichotomy appeared sufficient to introduce rules concerning the exploitation of resources and the interactions between the two populations. In this case, looking for parsimony did not lead to an impoverishment in terms of meaning from a thematic point of view.

In the enriched model, the choice has been made not to formalize "river" objects. In contrast to the forest/savannah dichotomy, implementing river objects was not consensual in the negotiation between modelers and thematic experts. It has been strongly suggested that rivers played an important role in the actual migration of human groups: not only did they provide additional resources with fishing, but they likely facilitated travel in dense forest environments. In a modeling framework carefully considering and integrating thematic knowledge, notably environmental data, it would thus have been necessary to take them into account (Davison et al. 2006). However, the formalization of rivers would have constituted a major, yet possible, change in the model. Creating a new object would have requested to specify new rules of interaction between groups and rivers, thus adding many parameters to the model with scarce data on which to base our choices. Instead, considering the enriched simulation model as an experimental laboratory, modelers defended the need to simplify the specification of the elements composing the environment. The key "explanatory" characteristic of rivers was the role they played in facilitating travel through the forest. This characteristic was shared with the "corridor" of savannah inside the forest, and the implementation in our model of the latter (see Fig. 6) thus actually also covered the former. While efficient in terms of modeling, this leads to a higher level of abstraction, adds an extra distance with the thematic field, and therefore requires careful interpretation of the results.

The differentiation between FF groups and Bantu groups gave rise to a co-constructed formalization which provided limited information on the diversity of human groups at various dates. For modelers, restricting the number of types of groups and of their attributes was a critical issue, since creating too many types or attributes would have required to formalize (too) many behavior rules, as well as to specify different interactions between groups of different types, or their differentiated interactions with the environment. The Bantu/FF dichotomy was in regard a fair and economical agreement between thematic experts and modelers, especially given the exploratory perspective adopted.

Despite our efforts, we made a number of modeling choices that likely oversimplified reality and should be reconsidered in more realistic simulations of Bantu/FF interactions. For instance, as for the differentiation between groups, we left aside the internal cultural diversity of both Bantu populations and FF populations. This indeed affected our ability to grasp the variability in the interactions between groups but seemed an acceptable simplification considering our modeling objectives and knowledge from the thematic experts.

Other modeling choices raise questions in regard to their validity throughout the entire simulation period. We, for example, decided that a group-agent never changed from one type to another (i.e., from FF type to Bantu type or the opposite) during the simulation. In other words, there is no phenomenon of "assimilation," something we justified by the fact that although the first encounters between Bantu and FF are likely more than 2,000 years old, genetic mixing only increased around 1,000 years ago (Coupé et al. 2017). Also, we did not model the progressive emergence of mixed populations along increasing interactions between FF and Bantu groups, a choice which again seemed well suited to our research goals but perhaps led our simulations to further depart from the past reality.

One should note that the types of the groups are not the only aspect that remains unchanged during the entire simulation. So are indeed the behaviors of the Bantu and FF groups. For instance, some thematic experts give credit to the hypothesis that the expansion of Bantu groups gradually had an effect on the presence of FF in the forest, leading FF groups to specialize in forest exploitation while savannah was increasingly occupied by Bantu groups. Such an evolution of behavior during the past was not taken into account in the model, again due to our modeling objectives, which did not involve the understanding of FF spatial dynamics.

A final choice to build rules generic enough to hold on over a period of a few thousands years was to rely on an "elastic" metaphor to flexibly qualify and formalize the link between Bantu and FF groups. It allowed us to account for interdependent movements without tackling the question of whether or not a form of domination existed between the two types of groups. This hypothesis of an elastic linkage that would remain identical at different periods of the simulation is very likely an oversimplification, given the expected evolution of the relations between Bantu and FF groups. However, the modalities of these changes were not clearly known by the thematic experts. Choosing the "elastic" formalization thus seemed rather generic, adapted to our modeling objectives, and overall suited to a large number of objects that coevolve in space.

Conclusion

In this chapter, we reported the co-construction of a multi-agent model to assess various hypotheses regarding the plausible routes followed by Bantu groups during their southward and eastward expansion, a large-scale migratory event which started approximately 4,000 years ago. In our view, from a modeling perspective, the co-construction has been successful. First and foremost, we succeeded in formalizing rules of behavior for the various agents. We were then able to evaluate the hypothesis that FF groups had an impact on the Bantu expansion. This was possible because of an in-depth dialogue between linguists, geographers, and modelers. On the one hand, the linguists benefited from this framework in that they could solidify a diffuse body of knowledge about the Bantu migrations and farmers' interactions with FF foragers – not only when looking at the final rules adopted but throughout the whole construction of the model. On the other hand, geographers and modelers developed their knowledge of mobility behaviors at time and space scales very remote from their habits. At this point, we have already worked on some experiments that were not presented in this chapter, and more detailed results are available in Coupé et al. (2017).

Turning now our attention to thematic outputs, our model exemplifies how the presence of FF groups, upon accepting a number of hypotheses, may have impacted the way Bantu groups crossed the forest. Interestingly, if the results of the simulation show that the presence of FF groups brings more Bantu groups to enter into the forest, they do not suggest that this presence either facilitated or impeded the crossing at a macroscopic level. However, the level of abstraction of the model does not allow us to really settle this complex thematic question. Additionally, our modeling framework has not yet produced stable thematic results, mainly because it is still at an exploratory stage. There might be a gap between linguists' expectations for the MAS approach before the start of the project and the results obtained so far: modeling is indeed a long process, especially in a pluridisciplinary context and with scarce data. More precisely, for instance, the model could not help thematic experts to discriminate between the "east-next-to-the-west" and "east-out-of-the-west" scenarios. We wish nevertheless to underline some outcomes of this process. Firstly, we were able to converge, rule by rule, on whether or not a behavior should hold during the entire simulation (e.g., the mutual attraction between Bantu groups and FF groups) or should evolve through time (e.g., the Bantu ability to use resources in the forest due to interaction with FF). Secondly, we were able to assess the genericity of the rules implemented in the model, hence the extent to which a rule is specific to the Bantu/FF case (e.g., a strong dissimilarity regarding the ability to extract forest resources; the asymmetry between Bantu and FF groups in terms of resources delivered to each other when the groups are paired by an "elastic link") or generic to the migrations of pre-agrarian or proto-agrarian human groups (e.g., the reification of groups with a stable size; the various motivations for a group to migrate; the variables involved in the choice of the next destination; the elastic linkage metaphor, etc.). It takes dialogue between thematic experts and modelers, and several adjustments in the model, to be able to converge on a shared vision of how to articulate thematic experts' knowledge, the research objectives, and the model itself, therefore seen as a co-constructed object.

Future work includes the implementation of a "genetic component" in the model, in order to compare the genes of the various groups at the end of the simulation and possibly relate them to recent (synchronous) empirical data. Indeed, even though the interaction rules that were implemented only apply to two kind of groups (Bantu and FF), the microscopic conformation of the MAS allows us to compute, along the simulation, the historicity of encounters between human groups and to some extent simulate genetic mixing, which outputs could be compared with actual data and help us to better evaluate plausible Bantu migratory routes. Overall, a further step would therefore be to scale up our model toward a more data-driven modeling framework, with in particular a fine-grained spatial description of the environment.

References

Alves, I., Coelho, M., Gignoux, C., Damasceno, A., Prista, A., & Rocha, J. (2011). Genetic homogeneity across Bantu-speaking groups from Mozambique and Angola challenges early split scenarios between East and West Bantu populations. *Human Biology, 83*(1), 13–38.

Ammerman, A. J., & Cavalli-Sforza, L. L. (1984). *The Neolithic transition and the genetics of populations in Europe*. Princeton: Princeton University Press.

Bahuchet, S. (1991). Les Pygmées d'aujourd'hui en Afrique centrale. *Journal des africanistes, 61*(1), 5–35.

Bahuchet, S. (2012). Changing language, remaining Pygmy. *Human Biology, 84*(1), 11–43.

Bahuchet, S., & Guillaume, H. (1982). Aka-farmer relations in the northwest Congo basin. In E. Leacock & R. Lee (Eds.), *Politics and history in band societies* (pp. 189–211). Cambridge: Cambridge University Press.

Barton, C. M., & Riel-Salvatore, J. (2012). Perception, interaction, and extinction: A reply to Premo. *Human Ecology, 40*(5), 797–801.

Barton, C. M., Riel-Salvatore, J., Anderies, J. M., & Popescu, G. (2011). Modeling human ecodynamics and biocultural interactions in the late Pleistocene of western Eurasia. *Human Ecology, 39*(6), 705–725.

Bastin, Y., Coupez, A., & De Halleux, B. (1983). Classification lexicostatistique des langues bantoues (214 relevés). *Bulletin des Séances de l'Académie Royale des Sciences d'Outre-mer, 27*(2), 173–199.

Bastin, Y., Coupez, A., & Mann, M. (1999). *Continuity and divergence in the Bantu languages: Perspectives from a lexicostatistic study*. Tervuren: Musée Royal de l'Afrique Centrale (MRAC), Annales, Série in-8°, Sciences humaines 162, 225 pp.

Biraben, J.-N., Masset, C., & Thillaud, P. L. (1997). Le peuplement préhistorique de l'Europe. In: *Histoire des populations de l'Europe, tome 1 – Des origines aux prémices de la révolution démographique*. Dupâquier J. and Bardet J.-P. (dir). Fayard, 660 p.

Bostoen, K. (2006). Pearl millet in early bantu speech communities in central Africa: A reconsideration of the lexical evidence. *Afrika und Übersee, 89*, 183–213.

Bostoen, K., Grollemund, R., & Muluwa, J. K. (2013). Climate-induced vegetation dynamics and the Bantu expansion: evidence from Bantu names for pioneer trees (*Elaeis guineensis*, Canarium schweinfurthii and Musanga cecropioides). *Comptes Rendus – Geoscience, 345*(7–8), 336–349.

Bostoen, K., Clist, B., Doumenge, C., Grollemund, R., Hombert, J.-M., Muluwa, J. K., & Maley, J. (2015). Middle to late Holocene paleoclimatic change and the early Bantu expansion in the rain forests of western central Africa. *Current Anthropology, 56*(3), 354–384.

Coupé, C., Hombert, J.-M., Le Néchet, F., Mathian, H., & Sanders, L. (2017). Transition 2: Modélisation de l'expansion des populations Bantu dans un espace déjà habité par des populations de chasseurs-collecteurs. In *Peupler la Terre: De la préhistoire à l'ère des métropoles*. Sanders L. (dir). Presses Universitaires François Rabelais, 527 p.

Currat, M., & Excoffier, L. (2005). The effect of the Neolithic expansion on European molecular diversity. *Proceedings of the Royal Society of London B: Biological Sciences, 272*(1564), 679–688.

Davison, K., Dolukhanov, P., Sarson, G., & Shukurov, A. (2006). The role of waterways in the spread of the Neolithic. *Journal of Archaeological Science, 33*, 641–652.

Edmonds, B., & Moss, S. (2004). From KISS to KIDS–an 'anti-simplistic' modelling approach. In *International workshop on multi-agent systems and agent-based simulation* (pp. 130–144). Berlin/Heidelberg: Springer.

Ehret, C. (1973). Patterns of Bantu and central Sudanic settlement in central and southern Africa (ca. 1000 B.C.–500 A.D.). *Transafrican Journal of History, 3*(1), 1–71.

Ehret, C. (2001). Bantu expansions: Re-envisioning a central problem of early African history. *The International Journal of African Historical Studies, 34*(1), 5–41.

Grollemund, R. (2012). *Nouvelles approches en classification: Application Aux Langues Bantu du nord-ouest*. Thèse de doctorat de l'Université Lumière Lyon 2.

Grollemund, R., Branford, S., Bostoen, K., Meade, A., Venditti, C., & Pagel, M. (2015). Bantu expansion shows Habitat alters the route and pace of human dispersals. *Proceedings of the National Academy of Sciences of the USA, 112*(43), 13296–13301.

Hassan, F. A. (1981). *Demographic archaeology*. New York: Academic Press.

Hazelwood, L., & Steele, J. (2004). Spatial dynamics of human dispersals; constraints on modelling and archaeological validation. *Journal of Archaeological Science, 31*, 669–679.

Heine, B., Hoff, H., & Vossen, R. (1977). Neuere ergebnisse zur territorialgeschichte der Bantu. In W. J. G. Möhlig, F. Rottland, & B. Heine (Eds.), *Zur sprachgeschichte und ethnohistorie in Afrika* (pp. 57–72). Berlin: Dietrich Reimer.

Henrici, A. (1973). Numerical classification of Bantu languages. *African Language Studies, 14*, 82–104.

Ichikawa, M. (2004). Mbuti. In R. B. Lee & R. Daly (Eds.), *The Cambridge encyclopedia of hunters and gatherers* (pp. 210–214). Cambridge: Cambridge University Press.

Kohler, T. A., Ortman, S. G., Grundtisch, K. E., Fitzpatrick, C. M., & Cole, S. M. (2014). The better angels of their nature: Declining violence through time among prehispanic farmers of the Pueblo southwest. *American Antiquity, 79*(3), 444–464.

Lake, M. W. (2014). Trends in archaeological simulation. *Journal of Archaeological Method and Theory, 21*, 258–287. Phan D. (dir), Ontologies et modélisation par SMA en SHS, Hermes-Lavoisier, Londres & Paris, 558 p.

Li, S., Schlebusch, C., & Jakobsson, M. (2014). Genetic variation reveals large-scale population expansion and migration during the expansion of Bantu-speaking peoples. *Proceedings of the Royal Society of London B (Biological Sciences), 281*, 20141448.

Maley, J. (2001). La destruction catastrophique des forêts d'Afrique centrale survenue il y a environ 2500 ans exerce encore une influence majeure sur la répartition actuelle des formations végétales. *Systematics and Geography of Plants, 71*, 777–796.

Neumann, K., Bostoen, K., Höhn, A., Kahlheber, S., Ngomanda, A., & Tchiengué, B. (2012). First farmers in the Central African rainforest: A view from southern Cameroon. *Quaternary International, 249*, 53–62.

Ngomanda, A., Neumann, K., Schweizer, A., & Maley, J. (2009). Seasonality change and the third millennium BP rainforest crisis in southern Cameroon (Central Africa). *Quaternary Research, 71*(3), 307–318.

Nonaka, E., & Holme, P. (2007). Agent-based model approach to optimal foraging in heterogeneous landscapes: Effects of patch clumpiness. *Ecography, 30*(6), 777–788.

Ortega, D., Ibáñez, J. J., Campos, D., Khalidi, L., Ménde, V., & Teira, L. (2016). Systems of interaction between the first sedentary villages in the Near East exposed using agent-based modelling of obsidian exchange. *Systems, 4*, 18. https://doi.org/10.3390/systems4020018.

Parisi, D., Antinucci, F., Natale, F., & Cecconi, F. (2008). Simulating the expansion of farming and the differentiation of European languages. In B. Laks (Ed.), *Origin and evolution of languages approaches, models, paradigms*. London: Equinox Publishing.

Patin, E., Siddle, K. J., Laval, G., Quach, H., Harmant, C., Becker, N., Froment, A., Régnault, B., Lemée, L., Gravel, S., Hombert, J.-M., Van der Veen, L., Dominy, N. J., Perry, G. H., Barreiro, L. B., Verdu, P., Heyer, E., & Quintana-Murci, L. (2014). The impact of agricultural emergence on the genetic history of African rainforest hunter-gatherers and agriculturalists. *Nature Communications, 5*, 3163.

Plaza, S., Salas, A., Calafell, F., Corte-Real, F., Bertranpetit, J., Carracedo, A., & Comas, D. (2004). Insights into the Western Bantu dispersal: mtDNA lineage analysis in Angola. *Human Genetics, 115*(5), 439–447.

Premo, L. S. (2006). Exploratory agent-based models: Towards an experimental ethnoarchaeology. In J. T. Clark & E. M. Hagemeister (Eds.), *Digital discovery: Exploring new frontiers in human heritage* (pp. 29–36). CAA, Computer applications and quantitative methods in archaeology, Proceedings of the 34th Conference, Fargo, USA, April 2006, Budapest, Archaeolingua, 2007.

Quintana-Murci, L., Quach, H., Harmant, C., Luca, F., Massonnet, B., Patin, E., Sica, L., Mouguiama-Daouda, P., Comas, D., Tzur, S., Balanovsky, O., Kidd, K. K., Kidd, J. R., van der Veen, L., Hombert, J.-M., Gessain, A., Verdu, P., Froment, A., Bahuchet, S., Heyer, E., Dausset, J., Salas, A., & Behar, D. M. (2008). Maternal traces of deep common ancestry and asymmetric gene flow between Pygmy hunter-gatherers and Bantu-speaking farmers. *Proceedings of the National Academy of Sciences of the USA, 105*(5), 1596–1601.

Rasteiro, R., Bouttier, P. A., Sousa, V. C., & Chikhi, L. (2012). Investigating sex-biased migration during the Neolithic transition in Europe, using an explicit spatial simulation framework. *Proceedings of the Royal Society B: Biological Sciences, 279*(1737), 2409–2416.

Russell, T., Silva, F., & Steele, J. (2014). Modelling the spread of farming in the Bantu-speaking regions of Africa: An archaeology-based phylogeography. *PLoS One, 9*(1), e87854.

Schwartz, D. (1992). Assèchement climatique vers 3000 B.P. et expansion Bantu en Afrique centrale atlantique : quelques réflexions. *Bulletin de la Société géologique de France, 3*, 353–361.

Vansina, J. (1995). New linguistic evidence and the Bantu expansion. *The Journal of African History, 36*(2), 173–195.

Young, D. (2002). A space-time computer simulation method for human migration. *American Anthropologist, 104*, 138–158.

From Past to Present: The Deep History of Kinship

Dwight W. Read

Introduction

The term "deep history" refers to historical accounts framed temporally not by the advent of a written record but by evolutionary events (Smail 2008; Shryock and Smail 2011). The presumption of deep history is that the events of today have a history that traces back beyond written history to events in the evolutionary past. For human kinship, though, even forming a history of kinship, let alone a deep history, remains problematic, given limited, relevant data (Trautman et al. 2011). With regard to a deep history, one conjecture is that human kinship evolved from primate social systems in a gradual, more-or-less continuous manner (see Chapais 2008); another conjecture is that kinship, in accordance with the incest account of Claude Lévi-Strauss (1969) or the fanciful, tetradic account of Nicholas J. Allen (1986), "comes into existence with a leap" (Trautman et al. 2011: 176); and yet another, the account to be developed in this paper, is that kinship, as it is understood and lived by culture bearers today, is the consequence of a profound and qualitative evolutionary transformation going from an ancestral primate-like social systems predicated on extensive face-to-face interaction to the relation-based social systems that characterize human societies (Read 2012).

A time depth for some of the aspects of kinship that are part of the deep history of kinship has been worked out, using data from hunter-gatherer societies. These data suggest that human kinship may have a deep history going back at least 50,000 years. Walker and co-workers (Walker et al. 2011) have concluded from phylogenetic reconstructions of marriage practices by hunter-gatherer groups that marriage has "a deep evolutionary history of limited polygyny and bride price/

D. W. Read (✉)
Department of Anthropology, University of California, Los Angeles, Los Angeles, CA, USA
e-mail: dread@anthro.ucla.edu

service that stems back to early modern humans and, in the case of arranged marriage, to at least the early migrations of modern humans out of Africa" (p. 2), dating to around 50,000 years BP. Likewise, Bancel and Matthey de l'Etang (2002) have argued that the so-called nursery kin terms—terms with a duplicated syllable formed from the sounds made by an infant, as in the English kinship address terms *mama* and *papa*—are candidates for being proto-kin terms in a reconstructed prototerminology due to nursery terms occurring worldwide with the same meaning. Matthey de l'Etang (2016) suggests that the widespread occurrence of the nursery term *kaka* found in Australian kinship terminologies, with meaning, mother's brother, implies that the hypothesized proto-kin term, denoted **kaka*, may have been brought to Australia by the first *Homo sapiens* reaching that continent, an event now dated to about 65,000 BP (Clarkson et al. 2017). This assumes *Homo sapiens* in Africa had already developed cultural systems that included the conceptual complexity of symbolic systems of kin terms that we refer to as kinship terminologies prior to the "out of Africa" migration. On the face of it, this seems unlikely. Instead, symbolic systems of kin terms organized in the form of kinship terminologies may first have been brought to Australia by a later Holocene migration of *Homo sapiens* from the Indian subcontinent dating to around 4000 BP (Pugach et al. 2013). In addition, **kaka* may refer to genealogical relations, rather than kin term relations, and the deep history of genealogical relations would almost certainly have a greater depth than the deep history of the symbolic system of kin terms we refer to as kinship terminologies, as will be discussed below.

In this paper I consider three formative, interconnected events in the deep history of the systems of kinship found in human societies. The first is the evolutionary beginning of kinship systems in human societies through biological evolution leading to the concept of a mother relation and of a father relation as part of hominin evolution leading to our species. The concepts of a mother relation and a father relation eventually coalesce through a cultural system of marriage that determines a spouse relation, forming what we can refer to as a cultural idea system (see Leaf and Read 2012) that conceptually links the father relation and the mother relation through the spouse relation into a single conceptually defined structure we will refer to as a Family Space. The second event, which develops interactively with the first through evolutionary changes occurring during hominin evolution leading to *Homo sapiens*, involves the development of the concept of genealogical connections, with origins that likely occurred more than 50,000 years ago, linking group members to one another through the logic of recursion applied to the mother relation and the father relation. The third is a monumental event that took place during the Upper Paleolithic when the earlier process of recursively tracing genealogical relations was transformed into a symbolic, computational system of kinship relations expressed through the kin terms making up a kinship terminology. This transformation was a remarkable intellectual achievement by our ancestors and provides the conceptual foundation for the kin term relations making up the kinship systems we find in human societies today. The goal of this paper is to show how these three events provide the foundation for an ontological account of the systems of kinship relations fundamental to human societies today.

Ontology of Kinship Relations

While the goal of this paper is to work out the deep history of kinship systems by relating events known from the paleontological evidence regarding changes that occurred during hominin evolution to each of the steps posited in an ontological account of kinship relations, what constitutes an ontological account for kinship relations in human societies needs to be clarified. Our understanding of the system of kinship relations expressed through kin terms—a system fundamental to understanding how the domain of kinship relations is structured and organized—has, from an analytical perspective, undergone a major transformation over the past several decades. The "received view" of an ontological account of the kinship relations central to all human societies traces back to the seminal work of Lewis Henry Morgan (1871) on the scientific analysis of kinship relations. This ontological account considers the foundation for kinship systems to begin with the nuclear family formed through marriage, followed by procreation as the source of the parent-child relations used to trace out genealogical relations connecting one kinsman to another. The account ends by assuming that kin terms are linguistic labels for categories of genealogical relations determined by largely unspecified behavioral and material factors (see Fig. 1).

However, the assumption that external factors determine the categories labelled by kin terms has been the Achilles' heel of this ontological account. As noted by Roy D'Andrade (2004: 311), the formal accounts of kinship relations developed in the 1960s and 1970s and based on this ontology are inadequate since "questions

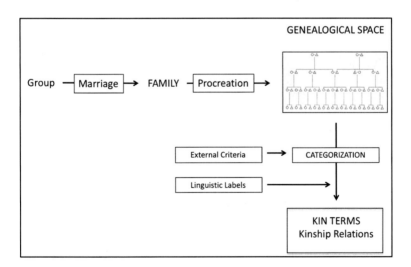

Fig. 1 Ontology for kinship relations according to the "received view." Genealogical relations are assumed to be the consequence of procreation, and kin terms are said to be linguistic labels for categories of genealogical relations formed through largely unspecified criteria external to the system of kinship relations

about why kinship structure took the forms they did were ignored." Crucially, this ontology, predicated upon genealogical relations determined through biological reproduction, has been argued to be ethnographically invalid by David Schneider (1984), leading him to proclaim "it is about time that we tested some other hypotheses" (Schneider 1972: 49). What a revised account corresponding to a different hypothesis would be, though, was left unstated by him.

Research work over the past several decades (see, e.g., Read 1984, 2001, 2007, 2010, 2015a, b, 2018a, b; Read and Behrens 1990; Leaf and Read 2012; Read et al. 2014) on the structural logic of kinship terminologies provides the revised account. This revised account is framed using ethnographic observations showing how kin terms, rather than playing a secondary role as linguistic labels for already established categories of genealogical relations as is assumed in the "received" view, form a logically coherent computational system that enables culture bearers to compute kin term relations directly from kin terms without reference to genealogical relations (let alone by reference to biological reproduction). The kin term computations used by culture bearers to do this is straightforward and can be illustrated with a simple example. English speakers, as culture bearers, easily make computations such as "If speaker refers to a person by the kin term uncle, and that person refers to another person by the kin term daughter, then speaker knows, based on his/her cultural knowledge, to refer to that other person by the kin term cousin." Or, more succinctly, cultural knowledge informs culture bearers that for their kin terms daughter, uncle, and cousin, "daughter of uncle is cousin" is a culturally valid expression showing the way the kin term cousin can be derived from the kin terms daughter and uncle. More formally, we can express this cultural knowledge regarding the kin term daughter, uncle, and cousin by observing that the pair of kin terms "daughter and uncle" in that order (i.e., considering the difference in meaning between the English kin term expressions "daughter of uncle" and "uncle of daughter") is mapped to the kin term cousin via cultural knowledge about the referential use of these kin terms. This computation may be expressed formally by the equation daughter o uncle = cousin, where the symbols "o" and "=" indicate that the pair of kin terms daughter and uncle is mapped to the kin term cousin in this computation.

More generally, we can use the equation, K o L = M (read: "K of L equals M"), as a way to express symbolically the following statement about the way the kin terms K, L, and M are conceptually linked: "If speaker refers to a person by the kin term L, and that person refers to another person by the kin term K, then speaker either knows, drawing upon his or her cultural knowledge, to refer to that other person by the kin term M or knows that there is no kin term referring to that other person." We will refer to this procedure by which a pair of kin terms K and L is mapped to a third kin term M as the *kin term product* of the kin terms K and L, in that order, and express symbolically the consequence for forming this product by the eq. K o L = M when, for culture bearers, there is a kin term M that speaker (properly) uses to refer to alter B if speaker refers to alter A by the kin term L and alter A refers to alter B by the kin term K.

The possibility that the product of a pair of kin terms may not be recognized by culture bearers as determining a kinship relation in their cultural repertoire can be seen, for example, in the English terminology with the kin terms parent and parent-in-law. If the speaker refers to alter A by the kin term parent-in-law and alter A refers to alter B by the kin term parent, then culture bearers would say that there is no kin term by which the speaker (properly) refers to alter B. Thus, while the kin term product of parent and parent-in-law is meaningful (i.e., the kin term product refers, meaningfully, to a situation where the speaker refers to alter A by the kin term parent-in-law and alter A refers to alter B by the kin term parent), the kin term product of parent and parent-in-law is not culturally recognized as a kin term relation (i.e., there is no English kin term by which the speaker properly refers to alter B in this situation), and so in the English kinship terminology, there is no kin term whose meaning would be expressed through the kin term product, parent of parent-in-law. For completeness of the formalism, we may express this possibility by the equation parent o parent-in-law = 0, where "0" means "not a kin term."

Some kin terms, M, are irreducible with respect to kin term products, meaning that there is no pair of kin terms K and L with K o L = M. For example, for English speakers, the kin terms mother and father are each irreducible since there is no pair of English kin terms K and L with K o L = mother or with K o L = father. We will refer to irreducible kin terms as *primary kin terms*. In English, the primary kin terms are the sex-marked consanguineal kin terms mother, father, son, daughter, husband, and wife and the neutral kin terms parent, child, and spouse that are cover terms for the sex-marked primary kin terms. The kin term sibling is not a primary kin term in the English terminology since child o parent = sibling. In other terminologies there may be sibling terms that are primary kin terms.

Another distinction regarding kin terms that is needed here is the distinction between ascending and descending primary terms, along with the fact that there may be a reciprocal kin term relationship between an ascending kin term and a descending kin term; e.g., in the English kinship terminology, mother and father are ascending primary kin terms with reciprocal kin terms son and daughter that are primary descending kin terms. In addition, some (but not all) terminologies will have sibling terms that are primary kin terms. The sibling terms in the English kinship terminology, as already noted, are not primary terms. Most, but not all, terminologies also have primary affinal terms such as husband, wife, and spouse for the English kinship terminology. For some terminologies, such as the Australian aboriginal terminologies (see Scheffler 1978), the term used to refer to one's spouse is one of the consanguineal kin terms (see Leaf and Read 2012; Denham 2013).

In the received view, the corpus of kin terms would only be structured in accordance with the logic for the classification of genealogical relations that are then linguistically labelled. In the revised ontology, the kin terms are structured by the logic of kin term products of primary kin terms, with the structure for the kin term products determined by structural equations that are part of the cultural repertoire for the group under consideration by stipulating the outcomes of kin term products for the pairs of kin terms that determine the structure of the kinship terminology. We can graphically represent the structure of a kinship terminology formed through taking kin term products of primary kin terms in the following manner. Let each kin term be a node in the structure for the terminology, and then connect the nodes by arrows representing products of primary kin terms with kin terms as follows. First, associate a different arrow form with each primary kin term so that the form of the arrow indicates which primary kin term is being used in a kin term product with the kin term located at a node in the structure. Second, draw the arrow corresponding to a primary kin term K from the kin term L located at a node to the node for the kin term M when the kin term product of the primary kin term K with the kin term L is the kin term M; that is, an arrow corresponding to the primary kin term K, beginning at the kin term L and ending at the kin term M, represents the kin term product eq. K o L = M, assuming the kin term product of K and L is recognized as being a kin term in the kinship terminology. Figure 2 shows the structure of the English kinship terminology and, for comparison, the structure of the Fanti (Ghana) kinship terminology (Kronenfeld 2009). Substantive differences in the structure of the two terminologies are immediately apparent just by visual comparison of the two structures.

The kin term map shows how kin terms form a structure through the use of kin term products of kin terms with primary kin terms. The form of the kin term map suggests that the kinship terminology can be generated from kin term products using the primary kin terms, beginning with products of the primary terms with themselves. Research over the past several decades has shown that a kinship terminology is formed in this manner through a generative logic expressing the kinship concepts held by culture bearers when they compute kinship relations directly from kin terms. The key point is that being able to generate the terminology without reference to genealogy invalidates the received view's assumption that kin terms are determined by categorization of genealogical relations shown in Fig. 1.

The last part of the revised ontology we need is the structure formed by the kinship relations expressed through the primary kin terms. This structure is fundamental to kinship relations since the primary kin terms and the structure that they form are, in a formal account, presumed to be self-evident to culture bearers, whereas

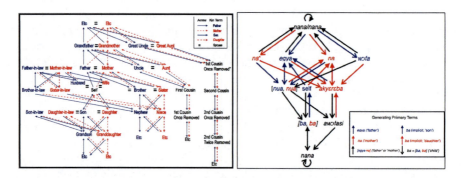

Fig. 2 (Left) Kin Term Space for the English kinship terminology, based on the primary kin terms father, mother, son, daughter, and spouse. In both diagrams, male terms are in blue, female terms are in red, and neutral terms are in black. Etc indicates that the structure continues in the same way without any structural changes. (Right) Kin Term Space for the Fanti (Ghana) terminology, based on the primary kin terms *egya* ("father"), *na* ("mother"), and *ba* ("child"). Affinal kin terms have not been included to make clearer the consanguineal structure. Structural differences between the two terminologies are visually obvious

From Past to Present: The Deep History of Kinship

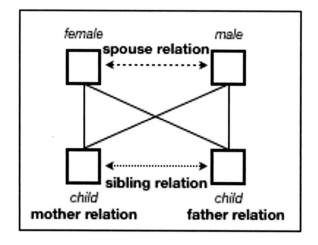

Fig. 3 Family Space. Solid lines indicate filiation. Boxes have content assigned through cultural instantiation

non-primary kin terms are defined through the kin term product, hence need not be self-evident without reference to the primary kin terms of the kinship terminology and its generative logic. Recognizing someone as a mother or a father, or reciprocally a son or daughter, is universal (though what constitutes the cultural criterion for being recognized as a mother or as a father is not universal) and understood by culture bearers without first needing to refer to a kinship terminology. The concept of being a spouse is also universal given that the cultural institution of marriage is, in some form, universal. The primary relations of mother and father, along with the reciprocal relations daughter and son, and the kin term relation spouse created culturally through marriage that conceptually links the mother relation to the father relation via the equations spouse of mother = father and spouse of father = mother can be expressed through the structure shown in Fig. 3. We will refer to this structure as a Family Space. The sibling relation between the two child positions in the Family Space is included as part of the Family Space, regardless of whether the sibling relation is a primary relation. We can now express the revised ontology in the following manner, using the Family Space, the Genealogical Space, and the Kin Space as the critical elements for that ontology (see Fig. 4). With this revised ontology in mind, we now relate the events of hominin evolution to the three critical elements of this ontology.

From Biologically Based Behavior to Cultural Relations

Individuation of Behavior Drives Primate Social Complexity

The first formative event, the conceptual formation of a Family Space of connections among group members, is an integral part of a major transformation that took place during hominin evolution leading to our species, *Homo sapiens*. The

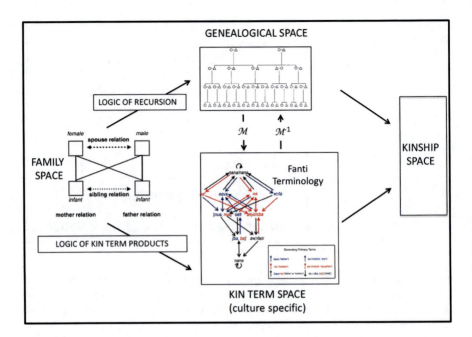

Fig. 4 Revised ontology for kinship relations. Family Space (see Fig. 3) is based on the mother relation, the father relation, the spouse relation, and the sibling relation. It provides the conceptual foundation for both the Genealogical Space and the Kin Term Space. The Genealogical Space is formed using the logic of recursion to compute genealogical pathways conceptually linking one individual to another using the relations making up the Family Space. The Kin Term Space is generated from the relations of the Family Space through the logic of kin term products, starting with products of the primary relations making up the Family Space. The Genealogical Space is mapped to the Kin Term Space by the mapping, M, defined by replacing each of the kin types in a genealogical pathway by its corresponding primary relation from the Family Space and then reducing this product of primary relations to a kin term in the Kin Term Space. The inverse of the mapping M, denoted by M^{-1}, determines the category of genealogical pathways corresponding to a kin term, K, by associating with K all of the genealogical pathways mapped to K by the mapping M. The categories of genealogical pathways associated with kin terms are predictable, and the predictions are found empirically to be 100% accurate

transformation goes from the phenomenal level of ancestral primate social systems based on face-to-face interaction for working out social relations among group members to the ideational level of relation-based systems of social interaction, for which culturally formulated kinship systems provide a canonical example, that characterize human societies. The evolutionary transformation leading to ideationally formulated systems of kinship relations made it possible for our species to accommodate, in a cohesive and coherent manner, the trend of increased individuated behavior that is part of a phylogenetic sequence going from the prosimians to the old-world monkeys and then to the great apes (Read 2012). This trend of individuated behavior would, if left unchecked, lead to cognitively unmanageable social complexity, especially with the introduction of the ability by chimpanzees to form male coalitions since social complexity correlates with the number of different behavior patterns that can occur in a group and for which each individual must learn to cope. Thus, social complexity will be proportional, at a minimum, to the number of distinct individuated behaviors expressed across group members plus the number of coalitions involving pairs of individuals each acting in accordance with either the same or a different individuated behavior that can be formed by group members. According to this measure, social complexity will be proportional to $n + (n/2)(n/2-1)/2 \sim n^2$, where n is the number of individuated behaviors (including both male and female behaviors); hence social complexity will increase with the square of the number of individuated behaviors when coalitions are also considered.

Formation of a Cognitive Constraint

What otherwise would become an unmanageable increase in social complexity as the number of individuated behaviors increased and as the behavioral formation of coalitions came into play was accommodated in primate societies in two ways. The first was through the adaptive expansion of cognitive abilities referred to by Robin Dunbar (1998) as the "social brain hypothesis" (see regression lines in Fig. 3). The second was through change in the form of social organization driven by the trend toward more individualized behavior (see content of ellipses in Fig. 5). The latter shifted social groups away from simple forms of social organization such as social behavior either being antagonistic or affiliative (as is the case for the prosimians [Jolly 1998: 5]; see Fig. 5, lemurs), toward the highly successful troop structures of the OW monkey (see Fig. 5, OW monkeys) that also included an increase in the size of social units, and then toward a chimpanzee fission-fusion form of social organization based on communities in which males form small, unstable social units and females are largely socially isolated (see Fig. 5, chimpanzees). Between the OW monkeys and the great apes, the size of social units decreased (but not the complexity of social interaction) as a way to cope with social complexity introduced by increased individuation of behavior, though the size of maximal agglomerations increased from the size of troops in OW monkeys (around 20–40 individuals) to the size of communities for the chimpanzees (from around 80 to 150 individuals). The chimpanzee communities likely came into play as the chimpanzees evolved into a species with a degree of individuated behavior comparable to that of humans (Yerkes 1927: 181; McGrew 2003: 179). This increased degree of individuated behavior, in combination with male-male coalitions, made, as discussed above, for a potentially power function increase in social complexity with the number of individuated behaviors in a group. This, in turn, cognitively limited the size of social units that could be achieved through biological kin selection alone in conjunction with face-to-face interaction as the means for establishing social interaction between individuated chimpanzees (Read 2012). The latter means that there was a cognitive barrier due to increased individuation of behavior potentially leading to overly complex social relations to which the chimpanzee communities adapted by reducing social complexity (see Fig. 6) through a structurally devolved social system

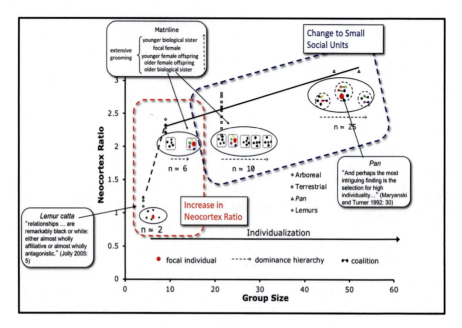

Fig. 5 Response to increase in social complexity in primate species due to increase in the individualization of behavior by increase in the neocortex ratio and reorganization of the structural organization of the primary social units. Prosimians, such as *Lemur catta* (lower left in the figure), have social relationships that are either affiliative or antagonistic, hence have social groups with social complexity $n \sim 2$ different behaviors (affiliative or antagonistic) by group members. Change from prosimians to the old-world monkeys involves a relatively large increase in the neocortex ratio (see dashed line). Old-world monkeys, divided into arboreal and terrestrial old-world monkeys, have a more complex social organization based on more individualistic behaviors, with females forming stable dominance hierarchies (see comment located in the upper left of the figure). Social complexity is based on individualistic behavior within a matriline and dominance relation between matrilines ($n \sim 6$–10). Increase in social complexity between the old-world monkeys and *Pan* is due to increase in individualistic behavior and the cognitive ability to form coalitions. Social complexity leads to socially solitaire females, and males only form unstable, small groups (up to 5–6 males). Male dominance hierarchy is unstable. Increase in neocortex ratio is relatively small, implying that coping with social complexity is largely through major changes in social organization in comparison to the OW monkeys

(in comparison to the OW monkeys) composed of communities of socially isolated females (Gagneux et al. 1999) and small, unstable male groups that are heavily dependent on face-to-face interaction and male-male grooming for even temporary social cohesion (Muller and Mitani 2005).

Expansion of STWM

Only recently in hominin evolution, and after extensive encephalization had already occurred by around 250,000 BP, did our ancestors have a sufficient increase in cognitive abilities derived from an increase in the size of short-term working memory

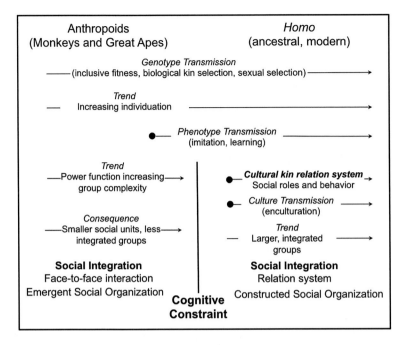

Fig. 6 Cognitive constraint. Solid discs: start of three new trends. (1) Phenotypic transmission, which provides the basis for traditions passed on from one generation to the next, traces back phylogenetically to the great apes and to some of the old-world monkeys. Trend of increasing individuation of behavior leads to a power function increase in social complexity when the formation of coalitions becomes part of the behavioral repertoire of chimpanzees. As shown in Fig. 5, trend of increasing social complexity leads to smaller social units and less integrated groups, thus to a cognitive constraint when face-to-face interaction is the basis for establishing social relations between individuated group members. (2) The cognitive constraint was circumvented non-biologically with the introduction of culturally formed social relation systems such as kinship systems. In social relation systems, social organization may be culturally formulated, hence "top down" rather than emergent. (3) Enculturation becomes the means for cultural transmission. (Redrawn from Fig. 4.5 in Read 2012)

(see below) to work out the beginnings of a cultural circumvention of the cognitive barrier faced by our chimpanzee-like primate ancestor (regarding what had previously been a cognitive barrier for biological adaptation alone, see Fig. 6). Our common primate ancestor with the chimpanzees would have had a short-term working memory limited to 2 ± 1 (Read 2008). A short-term working memory of this size is not large enough to work out the logic of recursion, which provides the logic for working out genealogical relations that can circumvent face-to-face interaction as the basis for social interaction to take place, since recursive reasoning requires a short-term working memory of size 3 at a minimum.

A critical change in cognitive abilities, then, relates to increase in the size of short-term working memory (STWM) that came into play during hominin evolution. The increase in STWM made possible cognitive complexity dependent upon cognitively taking into consideration several units of thought simultaneously.

For the chimpanzees, STWM = 2 ± 1 (Read 2008), which indicates that they are not able to consistently consider three or more concepts, ideas, or information units simultaneously.

We can track the increase in the size of STWM during hominin evolution through time-based changes in the design complexity of stone artifacts (see Fig. 7). The earliest stone artifacts made by hominins (corresponding taxonomically to *Australopithecus africanus*; see Fig. 7) consist of flakes removed from a nodule using the technique of conchoidal flaking. This would have required am increase to STWM = 3 since conchoidal flaking requires taking into consideration, simultaneously, the relationships among a hammer stone, the nodule from which the flake will be removed, and the angle of percussion required for conchoidal flaking. The cognitive complexity of

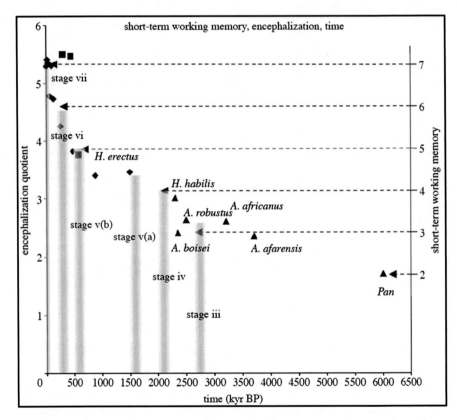

Fig. 7 Graph of encephalization quotient (EQ) estimates based on hominid fossils and *Pan* (chimpanzees). Early hominid fossils have been identified by taxon. Each data point is the mean for hominid fossils at that time period. Height of the "fuzzy" vertical bars is the hominid EQ corresponding to the data for the appearance of the stage represented by the fuzzy bar. Right vertical axis represents STWM. Encephalization data are adapted from the following: ▲ Epstein (2002); ■ Rightmire (2004); and ◆ Ruff et al. (1997). Phylogenetic groups for the encephalization data are identified except for the data for *Homo* post *H. erectus*. EQ = brain mass/(11.22 body mass 0.76). The stages refer to qualitatively different tool forms

conchoidal flaking is comparable to, or possibly exceeds that of, nut-cracking by chimpanzees as both require considering the relation among three objects: the anvil, the nut, and the hammerstone for nut-cracking and the nodule, the flaking angle, and the hammerstone for flake removal from a nodule. The kind of object manipulation involved in nut-cracking appears to be at the upper limit of the cognitive abilities of chimpanzees (Greenfield 1991; Parker and McKinney 1999), and about 25% of the adult chimpanzees never learn how to crack nuts (Biro et al. 2003) even though exposed, day in and day out, to chimpanzees that successfully crack nuts.

The next, qualitative design change are the Oldowan choppers that start appearing around 2.5 mya. They added an additional dimension that needs to be cognitively controlled, namely, flake removal, aimed at creating a sharp edge, conceptualized in one dimension as a line, thus requiring STWM = 4. STWM = 4 corresponds taxonomically to *Homo habilis* (see Fig. 7). The hand axes that first appear around 1.87 mya involve a shift from the one-dimensional line to the two-dimensional closed line that forms the boundary, and hence the shape, of a hand axe. The flaking of early hand axes focused primarily on the boundary of the hand axe and only to a minor extent on the surface bounded by the edge of the hand axe. The technology of the early hand axes involves, then, a shift from a single dimensional line to a two-dimensional closed curve, hence depends upon an increase of STWM to STWM = 5 (see Fig. 7). By around 500 kya, hand axes involve yet another design change with the flaking of the surface of the hand axe included in the technology of making hand axes. Though technologically more involved than is the case for the earlier hand axes, this may have only involved an increase in the average value of STWM, say to STWM = 5.5 on average (see Fig. 7). The next major qualitative change is the introduction of the Levallois technique for flake removal in which preparation of the core for the removal of a flake that, itself, will become the artifact and the repeated removal of flakes using the Levallois technique from the same core introduces yet another dimension to the technology of artifact production, namely, that of recursion as part of the technology of artifact production. This suggests STWM increases to STWM = 6, when the logic of recursion becomes part of the technology of artifact production (see Fig. 7). Lastly, the development of the blade technology involves both conceptualizing the core from which the blades will be removed in three dimensions and a fully recursive technology in which the removal of one blade prepares the core for the removal of the next blade. This corresponds to an increase in the size of working memory to STWM = 7 (see Fig. 7), that is, to a value of STWM comparable to that of modern *Homo sapiens*. With this as a background, we now turn to the three events making on the revised ontology and the evidence for their occurrence during hominin evolution leading to *Homo sapiens*.

Primary Relations and Evolution of the Family Space

The evolution of the Family Space does not correspond to a single evolutionary event but involves a broad and multifaceted transformation from behavioral actions taking place through processes carried out at the phenomenal level to a new framework in which behavioral actions now take place through the ideational level of a group's cultural idea systems that define and give shared meaning to actions undertaken by group members. The concept of a mother relation, for example, not only involves the concept of a connection between an offspring and the female who gave birth to the offspring but also incorporates assumptions about mothering behavior that is likely to be part of a mothering relation. The functionality derived through a concept like a mother relation does not stem simply from the mother relation as a trait at the individual level but as a group-level trait conceptualized in a comparable manner by group members; hence the shared concept provides synchrony across group members when behavior is formed in accordance with the concept of a mother relation.

The deep history of the cultural concept of a mother relation traces back, phylogenetically and pre-culturally, to old-world monkeys such as the macaques that categorize differences in the behavior of females to their own offspring and the behavior of females to the offspring of other females in one's troop (Dasser 1988). That behavior patterns of a female toward her own offspring differ from behavior patterns directed toward the offspring of other females follows directly from the fitness benefit obtained by a mammalian female with regard to mothering behavior directed by her toward her own offspring but not toward the offspring of other females. Though the historical depth of present-day macaque behavior is not known empirically, hence whether categorization similar to what occurs with present-day macaques also occurred in a common ancestor to the chimpanzees and the hominins is not known, categorization like this would certainly have been within the cognitive abilities of the hominins even early in their evolution leading to *Homo sapiens*.

The Concept of a Relation: The Mother Relation

While we do not have direct evidence for when and under what conditions the shift from categorization based on behavior patterns at the phenomenal level to the concept of a mother relation took place, we know that such a shift took place. The concept of a mother relation is universal in human societies, so we can assume that as part of the encephalization of the hominins that took place during their evolution leading to *Homo sapiens*, their cognitive ability expanded to the point where categorization at the phenomenal level of patterned behavior was transformed, through abstraction, into the concept of a mother relation linking a female to her offspring and, reciprocally, a child relation linking an offspring to her/his biological mother. Further, as communicative abilities among group members expanded, there would be convergence to a shared concept of a mother relation among group members due to similar behavior patterns by a female directed toward her offspring being the driver for the concept of a mother relation. Thus, part of the concept of a mother relation would be expected patterns of behavior by a mother toward her offspring. With this transition from categorization based on behavior occurring in response to events at the phenomenal level to categorization of a mother relation at the ideational level, we have one of the relations that subsequently becomes central to the concept of a Family Space.

The Relation of a Relation Is a Relation

A mother relation, in isolation, lacks the central feature of kinship relations, namely, the formation of a new relation through recursion through the idea that the relation of a relation is also a relation. Subsequent to the development of the concept of a mother relation, our hominin ancestors began to work out the concept that if B is recognized as having the mother relation to A, and C is recognized as having the mother relation to B, then C can be conceptualized as having a relation to A. The impetus for working out the idea that the relation of a relation is a relation may have a deep history tracing back to the events that took place during hominin evolution referred to in the "grandmother" hypothesis. The "grandmother" hypothesis posits that the uniquely human extension of female life expectancy beyond menopause relates to fitness benefits accruing to a female (through her daughter) when she extends her mothering behavior to her biological daughter's post-weaned offspring (Hawkes et al. 1998; Lahdenperä et al. 2004), thus expanding the scope of cooperative breeding across, and not just within, generations. The behavior posited for the grandmother hypothesis could have provided the impetus for recognizing a relation between the mother A of a mother B and the offspring of the mother B in the form of a "relation of a relation."

The conditions favoring accrual of fitness benefits by postmenopausal females engaging in the mothering behavior posited by the "grandmother" hypothesis would likely have come into play when there was an increased dependency on food resources that were difficult for weaned, but immature, offspring to obtain on their own, such as tubers, scavenged meat, or small hunted animals (O'Connell et al. 1999). This would also be the time frame for the occurrence of secondary altriciality, making females with offspring less mobile, which would also reinforce obtaining the potential fitness posited by the "grandmother" hypothesis. This also suggests that the extension of life expectancy beyond menopause may have begun as early as around 1–2 mya. If so, this would be the upper bound for when the concept that a relation of a relation is a relation became part of the cognitive repertoire of hominins ancestral to *Homo sapiens*. By 1 mya, STWM = 4–5 for the hominin ancestor to *Homo sapiens*, which is becoming large enough for reasoning about the relation of a relation being a relation. In addition, and critically, with the recognition that the mother of a mother is again a relation, the initial dependency of the mother relation being based on the biological mother is no longer central; that is, for the mother of a mother to be recognized as a relation, it suffices that person B is believed to be the mother of A and C is believed to be the mother of B for the mother of mother relation to be believed to hold between C and A, regardless of the biological connection between C and A.

The cognitive development of a mother relation concept along with the concept that a relation of a relation is a relation would have had far-reaching consequences. Once the concept that the relation of a relation is itself a relation is conceptually understood, it opens up forming chains of relations through the logic of recursion. Briefly, once female A conceptualizes the mother relation of female B to herself through the mothering behavior of B directed toward her, then female A may conceptually extend, through recursive reasoning, the mother relation to a female C who female A presumes to have the mother relation to female B. In this way, female A recognizes that female C is conceptually linked to her through recursive reasoning via "the mother relation of the mother relation." Implementation of recursive reasoning in this manner makes it possible to cognitively formulate a system of relations, such as the mother relation, the mother relation of the mother relation, the mother relation of the mother relation of the mother relation, and so on, from just the mother relation. In addition, and importantly, the concept of a mother relation also entails a reciprocal child relation that connects the offspring who is the target of the female engaging in mothering behaviors back to that female through affective behaviors directed by the offspring toward her. With the reciprocal child relation, the system of relations formulated through recursion may be expanded to include not only tracing back in time but forward from the past toward the present through recursively formed relations involving the child relation, such as the child relation of mother relation, the child relation of the mother relation of the mother relation, and so on.

Recursion and the Change from Individual-Level to Group-Level Traits

The system of relations constructed through recursion from the mother relation and its reciprocal child relation involves, it needs to be noted, more than just expansion of the number and variety of relations recognized as ways one individual may be linked to another. The mother relation, considered in isolation, does not involve a change in relationships among group members. Social relations among group members could still be predicated on face-to-face interaction as the primary means for working out social relationship among group members. With the introduction of recursive reasoning as a way to form new relations from already conceptualized relations, the transition from face-to-face interaction as the primary means to work out social relations to social interaction predicated upon already conceptualized social relations and associated, expected behavior patterns begins to play out. By working out a system of relations through recursion from, for example, the mother relation and its reciprocal child relation, the conceptual basis is thereby established by which one individual is not only able to conceptually recognize her connection with other females through the relations formed through recursion from the mother relation and the child relation, but the latter constitutes a system of relations understood in a common manner by group members. This provides the foundation for a profound shift in social relations. What is critical about a system of conceptually recognized relations is not the individual level of functional benefit that might be obtained for an individual to know her connections to other females in the group, but that by acting in accordance with expected behaviors associated with the system of relations, other group members will understand her behavior and her actions in a similar manner. In other words, there is a fundamental shift from a relation such as the mother relation seen in isolation, hence where the relation is similar to individually expressed traits, hence a situation where the trait's frequency is driven by the individual fitness derived from an individually expressed trait, to a system of traits where the system of traits becomes a group-level trait through the system of relations being understood by group members in the same manner and acted upon by group members in the same way. It is this transition from individual-level traits to group-level traits that is critical for the shift to relation-based system of social relations.

The Concept of the Father Relation

While recursive reasoning has the power to form a system of relations incorporating past and future time from just the mother relation and its reciprocal, the child relation, notably absent in this system built around the mother relation is the absence of the relation of a male to an offspring he has engendered. The reason for the absence is straightforward. From a biological perspective, there is no publicly observable biological marker for males for initiating a father relation comparable to pregnancy, birth, and nursing for females as the marker for initiating a mother relation. For this reason, male parenting is uncommon in the primates since, except in a few contexts, males cannot direct (implicitly or explicitly) parenting behavior toward their biological offspring. Male parenting likely came into play through changes that introduced secondary altriciality and otherwise restricted the mobility of females with newborn offspring. It would then have been in the interest of females to establish, at least temporary, emotional pair bonding with males as a means to induce provisioning by a male. Mathematical modeling shows that transition from promiscuous mating to pair bonding is unlikely even under widespread conditions (Gavrilets 2012). The modeling shows that conditions favoring this transition include increased female preference for provisioning coupled with difficulty of lower-ranked males getting access to females for mating purposes. These conditions would be met by the time of the Middle Paleolithic, when there were increased risk of female mortality when giving birth, increased need for high-quality resources such as meat, the risk of predation when females with altricial offspring scavenge (or hunt), and so on. These conditions would also increase competition among males for mating as the pool of sexually mature females would be reduced in relative size through increased female mortality associated with difficult births and scavenging and/or hunting; hence there would be increased selection for low-ranking males to gain mating access to females through provisioning, and simultaneously females would have an increased preference for being provisioned, thereby favoring selection for females who engage in behaviors (such as extended sexual receptivity) that ensure the regularity of provisioning by a male. Introduction of emotion-based pair bonding that characterizes *Homo sapiens* would appear to be the evolutionary consequence, and with emotion-based pair bonding, we can assume that male parenting behavior and the conceptualization of a father relation would be introduced into the cognitive repertoire of the hominin ancestors to *Homo sapiens*.

The Concept of a Spouse Relation

While emotional pair bonding may have sufficed to provide the conditions where there would have been biological selection for fathering behavior, emotional pair bonding still leaves uncertain the identification of biological father-child dyads by group members. Selection for male parenting does not require the correct identification of biological father-child dyads, only that male parenting behavior be sufficiently biased toward a male's biological offspring so that the net effect of male parenting behavior, even if directed occasionally toward non-biological offspring, or even if occasionally a male fails to engage in parenting behavior, is to increase positively the fitness of males. For a father-child relation to become part of the system of social relations coming into play through the system of relations built around the mother-child relation concept, the father-child relation must be recognized and identified by group members. In other words, for a father-child relation to become part of the system of relations built through recursion from a mother-child relation, the father-child relation must be identifiable with the kind of group public knowledge and certainty which the mother-child relation can be identified. Critical here is that the concern of group members is not in identifying the male who provided the sperm that impregnated which female but in a male being identified by the group the father of a child for the purpose of identifying the way in which group members are linked to one another through the logic of recursion acting on the mother relations (and its reciprocal) and the father relations; however it is determined by the group in a publicly agreed-upon manner.

The absence of a biological property that publicly identified a male as the father of an offspring in a manner comparable to the way a female is publicly identified as the mother of an offspring through pregnancy, birth, and subsequent mothering behavior had to be resolved in a non-biological manner. The absence of a biological criterion was solved culturally through public agreement, for a given female, on a male who will, for group social purposes, be presumed to be the father of any future offspring of that female (Chit Hlaing and Read 2016). We refer to the cultural assignment of a male as the presumed genitor of her offspring as a marriage between that male and that female (Malinowski 1929; Gough 1959), and we refer to the relation between them as a spouse relation. Phylogenetic evidence suggests that the cultural institution of marriage traces back 50,000 BP (Walker et al. 2011).

The combination of the mother relation, the father relation established through marriage, the spouse relation that specified the female whose offspring a male would be presumed to be the father, and the sibling relation conceptually linking the offspring of a female and, through marriage, a male, to one another, provides the conceptual foundation for the Family Space, hence the concept of a family as a fundamental social unit. Unlike the fission-fusion structure of unstable males units that are part of chimpanzee communities, family units formed in accordance with the relations making up the Family Space are structurally stable, coherent social units and provide a solid foundation as the social basis of a system of social organization.

Formation of a Genealogical Space

Developing a conceptual system of genealogical relations would not have been conceptually possible for our primate ancestors, as genealogical relations are based on the logic of recursion and recursive reasoning requires a short-term working memory of size 3 at a minimum. The evolutionary development of larger short-term working memory approaching the size of short-term memory in modern *Homo sapiens* made it possible to culturally circumvent the complexity of face-to-face social interactions due to the combination of individuated behavior and coalition formation that had formed a cognitive barrier for our chimpanzee-like ancestor. Our hominin ancestors were able to circumvent this barrier by abstracting from the phenomenal level of dyadic, patterned behavior through introducing, at the ideational level, the concept of a *relation* as a way to more abstractly characterize the connection between a pair of individuals interacting in a consistent, patterned manner (Read 2012).

The ensemble of the father relation culturally constructed through marriage in conjunction with the spouse relation, the mother relation initiated through birth, and their respective reciprocal relations jointly and conceptually form a Family Space (Read et al. 2014). The relations constituting the Family Space give rise to a system of genealogical relations through recursive reasoning. The genealogical relations make it possible not only for the members of a residence group of individuals living together on a day-to-day basis to conceptually and collectively formulate the relations they have to one another but to also formulate the connections the members of one residence group have to the members of another residence group. These connections are initiated through the pre-hominin and biological practice of individuals of one sex leaving one's natal group upon sexual maturity and joining another group for purposes of sexual reproduction but would now be augmented by marriage providing the cultural means for structurally incorporating the incoming individual through a spouse relation. In brief, a major transformation had now taken place in hominin deep history, qualitatively changing ancestral social systems based on face-to-face interaction to new forms of social systems based on culturally constructed systems of genealogical relations.

Limits of the Genealogical Space: Genealogical Complexity

Yet no society today has a kinship system based solely on genealogical relations. There are two major limitations recognized since the time of Lewis Henry Morgan (1871). First, the number of possible genealogical relations doubles in quantity with each increase in the number of parent-child steps used to trace genealogical relations. For the first step, there are eight possibilities, father, mother, son, daughter, brother, sister, husband, and wife; for the second step, there are $8 \times 8 = 64$ possibilities; for the third step, there are $8 \times 64 = 512$ possibilities; and so on. The system of genealogical relations rapidly becomes too extensive to be a coherent, consistent, stable, and mutually understood system of relations for representing the relations linking societal members by group members to one another, thus leading to a cognitive overload when trying to conceptualize the Genealogical Space as a whole (Lehman and Witz 1974; Chit Hlaing and Lehman 2011). The second problem is the difficulty in tracing genealogically to more distant collateral genealogical relations as this depends on the genealogical depth of parent-child tracing in the ascending direction, hence requires remembering genealogical pathways through individuals who are no longer living. The cognitive problems posed by these two cognitive limitations underscore why no human society has a system of kinship relations based solely on the Genealogical Space.

How the complexity of this system of genealogical relations was resolved brings us to the third kinship event, namely, the construction of a symbolic system for both expressing and computing kinship relations. The remarkable solution that our ancestors worked out to overcome these cognitive barriers was to abstract from the concatenation of genealogical relations the concept of a product that we refer to as a kin term product, defined over Family Space relations viewed as a system of symbols for which a product of symbols was definable in which the symbol product, unlike concatenation, is neither determined simply by the form of the symbols nor by their instantiation but through their usage in expressing the (conceptual) relation of one individual to another. Further, the form and properties of the structure generated through kin term products are determined by structural equations expressing cultural concepts that a system of kinship relations expressed in this manner should satisfy, such as the universal notion of the reciprocity of kinship relations.

Kin Term Space and Kinship Terminologies

What our ancestors achieved by working out a system of kin terms generated using the kin term product is, then, truly remarkable. Even assuming the concept of a Genealogical Space and the relative product of genealogical relations were already worked out, to work out a system of kin terms, they still had to abstract from the Genealogical Space an internally consistent conceptually closed system of kinship terms as a way to express kinship relations, with the system of kin relations small in size (kinship terminologies typically have around 15–25 kin terms) and with kin term relations computed using the kin term product through which a Kin Term Space is generated. As if this were not enough, by forming the Kin Term Space from the relations making up the Family Space, a simple mapping links genealogical relations onto kin terms, as already discussed; hence the Kin Term Space incorporates the Genealogical Space in the sense that there is an associated kin term for each genealogical position and, conversely, each kin term in the Kin Term Space determines a category of genealogical relations. This was, to say the least, a stupendous achievement.

Evidence for the Formation of a Kin Term Space: Upper Paleolithic

We have evidence—monumental in form, as is appropriate—for when this incredible achievement took place and what were involved. Relevant here is Claude Lévi-Strauss's (1962: 128) insight that hunter-gatherers express the ordering of their social universe to themselves through animal species: "Les espèces sont choisies non commes bonnes à manger, mais comme bonnes à penser" ("We can understand, too, that natural species are chosen not because they are 'good to eat' but because they are 'good to think'." [translation by E. Leach]). The "thinking through animals" occurs in Chauvet Cave, France, dating to 35,000 BP, through the content and organization of the animal depictions covering the cave walls (Leaf and Read 2012). These incredible depictions are not a literal representation of what artisans saw around them. The images are strikingly realistic, showing the features of animals in detail, as if the intent is to focus on the way even animals of the same species have individuality, yet at the same time the groupings of animals are by kinds, yet not in groups that are not seen in reality. Thus, there is the depiction of four horses, yet each with the physical attributes of a horse during one of the four seasons. In brief, the paintings represent the individuality of animals, yet they are arranged and organized in a manner that is a creation of the mind. What the arrangement and organization of animals painted on the cave walls depicts, over and over again, is an opposition between individuality expressed through an animal's unique features and a collective identity depicted through groups of animals of the same kind. The artisans and their audience know from their experience as hunters that one animal kind is interconnected with different animal kinds; thus animals form a collective whole despite differences in kind. Individual animals, kinds of animals, and their interconnectedness in the form of collectively all being animals are a way to think about and to express a new, and profound, notion of transforming the space of individual genealogical relations that expresses how the individuality of those who are members of the same living group forms a collectivity of individuals of the same kind, that is, sharing the property of being genealogically connected to each other, thus conceptually making the group of those living together into a coherent whole despite the individuality of each member of that group. At the same time, living groups can be connected to other living groups, despite the differences between living groups that previously kept them apart in time and space, like the Neanderthal living groups apparently without connection among living groups, as indicated by DNA evidence showing that Neanderthal mating was within the same living group of Neanderthals.

What the *Homo sapiens*, living at the same time and in the same region, were working out and displaying through the depictions of individual animals grouped into a single kind, yet sharing commonality at another level through all being animals, was the profound notion that the system of genealogically grounded relations could be transformed into a system of relations generated and organized symbolically through which different groups were now conceptualized as forming a single system of families connected conceptually through a symbolic system of kinship relations. In effect, they were displaying on the walls of Chauvet Cave the remarkable idea that just as the kinds of animals depicted in Chauvet Cave constitute a whole, the individuated categorization of the genealogical relations giving coherence to a single living also could be seen as constituting a whole expressed through a closed and bounded symbolically generated conceptual system of kinship relations making up the concept of a Kin Term Space in which the particular form and scope of this space were shaped by the ideas they had about the kinship relations connecting the individuals in one living group to another living group, thereby conceptually forming a single collectivity. The transformation from knowing how individuals of a living group form a whole through parent-child connections, which was not bounded in its scope but had the limitation of becoming non-comprehensible as the length of the chain of genealogical connections increases, to a coherent system of encompassing relations constructed symbolically from the same parent-child relations was one of the most profound transformations in human history, and the monumentality of this transformation could only be conveyed to those living in different living groups, previously isolated in time and space, through the monumental event of depicting these ideas through the organization and structure imposed on the kinds of animals depicted on the walls of Chauvet Cave.

Yet this poses a problem. Chauvet Cave dates to 35,000 BP, whereas the "out of Africa" migration occurred 30,000 years earlier, meaning that the kinship concepts those migrants would have had were not symbolic systems of kinship terms but concepts about kinship relations limited to some form of genealogical tracing. Living groups would still be more-or-less isolated from one another at the time of this migration. In this migration, two land masses, previously isolated from hominin evolution, were now involved, one early on and the later much later. For the later migration, *Homo sapiens* first reached North America around 15 kya, but by this time the systems of kinship relations worked in contexts such as Chauvet Cave would already be part of the cultural adaptation of the migrants, and what we see in the societies that developed in the Americas is the cultural equivalent of biological adaptive radiation. The independent evolution of kinship system in the context of the Americas that parallels the development of kinship systems in other parts of the world attests to structural constraints limiting the number of different logically coherent systems of kin terms that could be put together; hence convergent evolution must have been common.

The other land mass is Australia, where migrants arrived around 65 kya. Here, migration appears to be a one-time event prior to European contact with Australia, with the exception of a migration from the Indian subcontinent to Australia that occurred about 4 kya. This poses the question: Do the kinship systems that characterize the Australian groups at the time of European contact derive from the earlier migration that took place around 65 kya, or were these kinship systems brought to Australia with the migration from the Indian subcontinent? If the former, then there must have been two independent "inventions" of symbolic systems of kinship, one due to those living in the region around Chauvet Cave and the other due to groups in Australia (and possibly elsewhere). This implies that similarities between kinship systems found in Australia with kinship systems found elsewhere are the consequence of independent "invention" of symbolic kinship systems. Further, it raises a problem with the assumption that the kinship systems of Australia are the ancient form from which other forms of kinship evolved, for just as Australia was isolated from in-migration, it was also isolated from out-migration before about 4 kya. Alternatively, and less problematic, it is possible that the groups in Australia did not have symbolic systems of kinship relations until around 4 kya with the migration from the India subcontinent to Australia.

Conclusion

A deep history of kinship, though necessarily speculative in its details, carries with it the potential of bringing out the profound transformations leading from the phenomenal level of social systems based on face-to-face interaction that the beginnings of hominin evolution share with our primate ancestors to the ideational level at which the concepts make up the kinship systems that characterize human societies and are given concreteness, or as Roy Wagner (2016) has put it, through the concept of a system of symbolic kin terms expressing how individuals and families are connected to one another whether as part of the same or as part of different living groups. The shift from social systems based on the phenomenal level of face-to-face interaction to the ideational level of relation-based systems of social organization was not an elaboration on what was already present, even if only in nascent form, in our primate ancestors, but involved a profound transformation that made us a species whose adaptation became dependent upon the formation and transmission across generations of a shared system of ideas we refer to as culture and for which the symbolic systems of kinship relations through kin terms are the canonical form. Despite the centrality of kinship systems for human societies, and especially for small-scale, hunter-gatherer societies, the evolutionary sequence leading to the formation of a concept of kinship relations expressed through a symbolic system of kin terms through which kinship relations among individuals may be identified has, for the most part, been outside of current evolutionary stories of hominin evolution. Working out a deep history of kinship systems is a way to address this lacuna.

References

Allen, N. J. (1986). Tetradic theory: An approach to kinship. *Journal of the Anthropological Society of Oxford, 17,* 87–109.

Bancel, P. J., & Matthey de l'Etang, A. (2002). Tracing the ancestral kinship system. The global etymon *kaka*. Part I: A linguistic study. *Mother Tongue, 7,* 209–243.

Biro, D., Inoue-Nakamura, N., Tonooka, R., Yamakoshi, G., Sousa, C., & Matsuzawa, T. (2003). Cultural innovation and transmission of tool use in wild chimpanzees: Evidence from field experiments. *Animal Cognition, 6,* 213–223.

Chapais, B. (2008). *Primeval kinship: How pair-bonding gave birth to human societies.* Cambridge: Harvard University Press.

Chit Hlaing, F. K. L. (Lehman, K.). 2011. Kinship theory and cognitive theory in anthropology. In D. B. Kronenfeld, G. Bennardo, V. C. de Munck and M. D. Fischer (eds.) A companion to cognitive anthropology. Pp. 254–269. London: Blackwell Publishing Ltd.

Chit Hlaing, F. K. L., & Read, D. W. (2016). Why marriage? *Structure and Dynamics: eJournal of Anthropological and Related Sciences, 9*(2). Retrieved from http://escholarship.org/uc/item/56b9b0rb.

Clarkson, C., Jacobs, Z., Marwick, B., Fullagar, R., Wallis, L., Smith, M., Roberts, R. G., Hayes, E., Lowe, K., Carah, X., Florin, S. A., McNeil, J., Cox, D., Arnold, L. J., Hua, Q., Huntley, J., Brand, H. E. A., Manne, T., Fairbairn, A., Shulmeister, J., Lyle, L., Salinas, M., Page, M., Connell, K., Park, G., Norman, K., Murphy, T., & Pardoe, C. (2017). Human occupation of northern Australia by 65,000 years ago. *Nature, 547,* 306–310.

D'Andrade, R. (2004). Why not cheer? *Journal of Cognition and Culture, 3*(4), 310–314.

Dasser, V. (1988). A social concept in Java monkey. *Animal Behaviour, 36*(1), 225–230.

Denham, W. W. (2013). Beyond Fictions of Closure in Australian Aboriginal Kinship. *Mathematical Anthropology and Culture Theory, 5*(1), 1–90. Retrieved from https://escholarship.org/uc/item/7d69w4sk.

Dunbar, R. I. M. (1998). The social brain hypothesis. *Evolutionary Anthropology: Issues, News, and Reviews, 6*(5), 178–190.

Epstein, H. T. (2002). Evolution of the reasoning brain. *Behavior and Brain Science, 25,* 408–409. https://doi.org/10.1017/S0140525X02270077.

Gagneux, P., Boesch, C. & Woodruff, D. S. (1999). Female reproductive strategies, paternity and community structure in wild West African chimpanzees. Animal Behaviour, 57, 19–32

Gavrilets, S. (2012). Human origins and the transition from promiscuity to pair-bonding. *Proceedings of the National Academy of Sciences USA, 109,* 9923–9928.

Gough, E. K. (1959). The Nayars and the definition of marriage. *The Journal of the Royal Anthropological Institute of Great Britain and Ireland, 89*(1), 23–34.

Greenfield, P. (1991). Language, tools and brain: The ontogeny and phylogeny of hierarchically organized sequential behavior. *Behavioral and Brain Sciences, 14,* 531–595.

Hawkes, K., O'Connell, J. F., Blurton Jones, N. G., Alvarez, H., & Charnov, E. L. (1998). Grandmothering, menopause, and the evolution of human life histories. *Proceedings of the National Academy of Sciences USA, 95,* 1336–1339.

Jolly, A. (1998). Lemur social structure. *Folia Primatologica, 69*(suppl.1), 1–13.

Kronenfeld, D. (2009). *Fanti kinship and the analysis of kinship terminologies.* Urbana: University of Illinois Press.

Lahdenperä, M., Lummaa, V., Helle, S., Tremblay, M., & Russell, A. F. (2004). Fitness benefits of prolonged post-reproductive lifespan in women. *Nature, 428*(6979), 178–181.

Leaf, M., & Read, D. (2012). *The conceptual foundation of human society and thought: Anthropology on a new plane.* Lanham: Lexington Books.

Lehman, F. K., & Witz, K. (1974). Prolegomena to a formal theory of kinship. In P. Ballonoff, *Genealogical mathematics* (pp. 111–134). Paris: Mouton.

Lévi-Strauss, C. (1962). *La Pensée Sauvage.* Paris: Plon.

Lévi-Strauss, C. (1969[1949]). *The elementary structures of kinship* (trans: Weightman, J., & Weightman, D.). New York: Harper & Row.
Malinowski, B. (1929). Marriage. *Encyclopedia Britannica, 14*, 940–950.
Matthey de l'Etang, A. (2016). Kv(Ŋ)Kv- Kinship terms in the Australian Aboriginal languages. First Part: Kaka 'Mother's Brother'. Structure and Dynamics, 9(2). Retrieved from https://escholarship.org/uc/item/61z81220
McGrew, W. C. (2003). *The cultured chimpanzee*. Cambridge: Cambridge University Press.
Morgan, L. H. (1871). *Systems of consanguinity and affinity in the human family*. Washington, D.C.: Smithsonian Institute.
Muller, M. N., & Mitani, J. C. (2005). Conflict and cooperation in wild chimpanzees. *Advances in the Study of Behavior, 35*, 275–331.
O'Connell, J. F., Hawkes, K., & Blurton Jones, N. G. (1999). Grandmothering and the evolution of Homo erectus. *Journal of Human Evolution, 36*(5), 461–485.
Parker, S. T., & McKinney, M. L. (1999). *The evolution of cognitive development in monkeys, apes, and humans*. Baltimore: The Johns Hopkins University Press.
Pugach, I., Delfin, F., Gunnarsdóttir, E., Kayser, M., & Stoneking, M. (2013). Genome-wide data substantiate Holocene gene flow from India to Australia. *Proceedings of the National Academy of Sciences (USA), 110*(5), 1803–1808.
Read, D. (1984). An algebraic account of the American kinship terminology. *Current Anthropology, 25*, 417–440.
Read, D. (2001). What is kinship? In R. Feinberg & M. Ottenheimer (Eds.), *The cultural analysis of kinship: The legacy of David Schneider and its implications for anthropological relativism* (pp. 78–117). Urbana: University of Illinois Press.
Read, D. (2007). Kinship theory: A paradigm shift. *Ethnology, 46*(4), 329–364.
Read, D. (2008). Working memory: A cognitive limit to non-human primate recursive thinking prior to hominid evolution. *Evolutionary Psychology, 6*(4), 603–638.
Read, D. (2010). The generative logic of Dravidian language terminologies. *Mathematical Anthropology and Cultural Theory, 3*(7). http://www.mathematicalanthropology.org/pdf/Read.0810.pdf. Accessed 24 Sept 2010.
Read, D. (2012). *How culture makes us human* (Series: Big ideas in little books). Walnut Creek: Left Coast Press.
Read, D. (2015a). Formal models of kinship. In J. D. Wright (editor-in-chief), *International encyclopedia of the social & behavioral sciences* (Vol. 13, 2nd ed., pp. 53–60). Oxford: Elsevier.
Read, D. (2015b). Kinship Terminology. In J. D. Wright (editor-in-chief), *International encyclopedia of the social & behavioral sciences* (Vol. 13, 2nd ed., pp. 61–66). Oxford: Elsevier.
Read, D. (2018a). The extension problem: Resolution through an unexpected source. In W. Shapiro (Ed.), *Focality and extension in kinship: Essays in memory of Harold W Scheffler* (pp. 59–112). Melbourne: Australian National University.
Read, D. (2018b). The generative logic of Crow-Omaha terminologies: The Thonga-Ronga kinship terminology as a case study. *Mathematical Anthropology and Culture Theory, 12*(1), 1–38.
Read, D., & Behrens, C. (1990). KAES: An expert system for the algebraic analysis of kinship terminologies. *Journal of Quantitative Anthropology, 2*, 353–393.
Read, D., Fischer, M. D., & Lehman, F. K. (2014). The cultural grounding of kinship: A paradigm shift. *L'Homme, 210*, 63–89.
Rightmire, G. P. (2004). Brain size and encephalization in Early to Mid-Pleistocene Homo. *American Journal of Physical Anthropology, 124*, 109–123. https://doi.org/10.1002/ajpa.10346.
Ruff, C. B., Trinkhaus, E., & Holliday, T. W. (1997). Body mass and encephalization in Pleistocene Homo. *Nature, 387*, 173–176. https://doi.org/10.1038/387173a0.
Scheffler, H. (1978). *Australian kin classification* (Cambridge studies in social anthropology no. 23). Cambridge: Cambridge University Press. https://doi.org/10.1017/CBO9780511557590.
Schneider, D. M. (1972). What is kinship all about? In P. Reining (Ed.), *Kinship studies in the Morgan centennial year* (pp. 32–63). Washington, D.C.: The Anthropological Society of Washington.

Schneider, D. (1984). *A critique of the study of kinship*. Ann Arbor: University of Michigan Press.

Shryock, A., & Smail, D. L. (Eds.). (2011). *Deep history: The architecture of past and present*. Berkeley: University of California Press.

Smail, D. L. (2008). *On deep history and the brain*. Berkeley: University of California Press.

Trautman, T. R., Feeley-Harnik, G., & Mitani, J. C. (2011). Deep kinship. In *Deep history: The architecture of past and present* (pp. 160–190). Berkeley: University of California Press.

Wagner, R. (2016). The nexus between kinship and ritual. *Structure and Dynamics, 9*(2), 240–251. https://escholarship.org/uc/item/7rq8w3ff.

Walker, R. S., Hill, K. R., Flinn, M. V., & Ellsworth, R. M. (2011). Evolutionary history of hunter-gatherer marriage practices. *PLoS One, 6*(4), e19066. https://doi.org/10.1371/journal.pone.0019066.

Yerkes, R. M. (1927). A program of anthropoid research. *American Journal of Psychology, 34*, 181–199.

Modeling the Relational Structure of Ancient Societies through the *Chaîne opératoire*: The Late Chalcolithic Societies of the Southern Levant as a Case Study

Valentine Roux

Introduction

In archeology, modeling evolution processes is a major issue. The goal is to explain cultural variation over space and time in light of general evolutionary mechanisms – inheritance, interaction, and local adaptation (Shennan et al. 2015). Modeling these processes is done using different tools (computational models, simulations; for a review, see Cegielski and Rogers 2016). General evolutionary models exploring cultural transmission for explaining changes in material culture through time are mainly found in evolutionary and network archeology. As a way to introduce the methodological questions dealt with in this paper, I first briefly recall what these two modeling approaches aim at.

In evolutionary archeology, models aim at interpreting artifact changes in terms of mode of transmission (number of people involved, direction in which information is passed – phylogenesis *versus* ethnogenesis – biased forms of transmission, and how information is packaged; Eerkens and Lipo 2007). Studying modes of transmission implies to understand how information is acquired and transformed by individuals through microevolutionary process (selection, mutation, drift). These models quantify the effects of sociopsychological mechanisms in time (e.g., copying the most prestigious, conforming to the majority, copying with errors; Mesoudi 2009) and offer reference patterns of variability depending on the modes of transmission (e.g., Bentley and Shennan 2003; Bettinger and Eerkens 1999; Jordan and Shennan 2003; Shennan and Wilkinson 2001; Tehrani and Collard 2009). They thus aspire to explain the large-scale patterns observed in the archeological record by extrapolating microevolutionary processes in time and space (Mesoudi and O'Brien 2009) and investigating the spatial and temporal structure of cultural variation

---

V. Roux (✉)
CNRS, UMR 7055, University Paris-Nanterre, Nanterre, France
e-mail: valentine.roux@cnrs.fr

(Shennan et al. 2015). Although transmission of information is acknowledged to occur at the level of the group (Mesoudi and O'Brien 2009), the context into which information is transmitted is less explored (except for the size of the population; O'Brien and Bentley 2011; Powell et al. 2010; Shennan 2001).

These evolutionary models raise several questions. One of them is the nature of the traits used in these models. They are mainly morphocentric (i.e., modeling evolution of morphometrical traits) (Dunnell 1978; Eerkens and Lipo 2005; Gandon et al. 2014; Hamilton and Buchanan 2009; Neiman 1995; O'Brien et al. 2010; Shennan and Wilkinson 2001) and ignore technological traditions which can be defined as inherited ways of doing things and which require to be described with other traits (Charbonneau 2018). Secondly interpretation of the evolution of traits is reduced to elementary mechanisms of change (e.g., interpreting evolution in terms of types of learning – unbiased transmission *versus* bias transmission), whereas these elementary mechanisms are not sociological regularities (specifying the social structure within which these mechanisms operate). As such, they can measure the change (its direction, tempo, and scale), but cannot be used for explaining why certain social structures are more favorable than others to evolution processes (e.g., why cultural diffusion occurs in some places rather than in others) and therefore for bringing to light potential evolutionary laws explained by social facts that we could use to interpret evolutionary processes (Gallay 2011).

In contrast, analytical sociology focuses mainly on the relational structure of societies, namely, the social network, within which information is transmitted and diffused (Axelrod 1997; Valente 1999; Valente 1996). Network models are used to relate individual actions (micro-level), the interdependence structures between those actions (the interactions between the units, meso-level), and the sociological regularities emerging from the latter (macro-social generalizations) (Manzo 2007, 2014). These models use simulation methods, including multi-agent system.[1] The ambition of this method is to unveil the mechanisms explaining how regularities are created, knowing that it is not enough to produce a result for claiming that the activated individual actions are the explanatory factors. For this reason individual actions are considered in relationship with different types of interactions (the interdependence structures) in order to understand better how some individual actions can generate macro-social regularities. Thus for the purpose of explaining the social conditions favorable to innovation or diffusion of cultural traits, individual actions are simulated within different network structures (i.e., homogeneous *versus* heterogeneous; Flache 2018; Flache and Macy 2011; Granovetter 1983; Rogers 1962). In these models, the content of the information is also measured (Rogers 1962), the spread of information considered as depending on both the network structures and the content of what is being transmitted (Centola 2015; Centola and Macy 2007).

---

[1] "a multi-agent system is made up of a set of n elementary units (named 'automata' or 'agents'). The researcher can program both the behavior of these units, either singly or grouped into subsets, and the way the units (or group of units) interact in time. The aim of the technique is to observe how the system of interaction between agents evolve and its final 'emerging' configuration" (Manzo 2007, p. 49).

Like the models used in evolutionary archeology, simulations from analytical sociology are based on individual actions grounded in sociopsychological rules. However, because these individual actions are considered in different contexts of interactions and thus the meso-level modeled for understanding the micro-macro problem (Manzo 2007: 51), the regularities produced offer hypotheses about the social structures favorable to changes. These regularities are looked for in archeology because they provide explanations to correlates between social structure and changes, knowing that explanatory mechanisms underlying these correlates cannot be studied in archeology given the lacunar aspect of the documentation. In other words, archeological data may allow us to describe processes of change in terms of mode of transmission (e.g., ethnogenesis *versus* phylogenesis); however they do not allow us to test why a specific social structure was favorable to change (e.g., why it favored or not social influence and led or not to assimilation). Such a test can be done only based on actualist data, no matter the computational tools used, because explanatory mechanisms refer to individual actions that cannot be explored in the past.

In light of this, understanding cultural processes by reference to sociological regularities raises a major issue: the characterization of ancient network structures. This characterization is necessary for a comparison with simulated network structures in order thereafter to benefit, by analogy, of the regularities associating network structure and process of change (e.g., relationship between weak ties[2] and diffusion) and the explanations given to the role of social structure in the processes of change (e.g., why weak ties favor diffusion). In archeology, network analysis has been applied mainly to reconstruct ancient interregional connection networks (Brughmans 2010, 2013; Collar et al. 2015; Knappett 2011; Östborn and Gerding 2014). Local networks and therefore the relational structure of societies are less studied even though they are acknowledged to be determinant for understanding evolutionary phenomena (Blake 2014; Knappett 2018). This is partly due to difficulties in finding relevant proxies for inferring social relationships between sites. The same issue applies when investigating cultural groups or phylogenetic links (Perlès 2013; Shennan et al. 2015).

In this respect, the interpretation of archeological data, whether using computational or simulation tools, faces two main methodological issues, discussed here: (1) the variables for expressing the relational structure of a society and (2) the use of sociological models for explaining evolution process, taking account of the lacunar and polysemic nature of archeological data.

Both issues are discussed in light of an archeological case study which raises the question of the social context wherein new ceremonial objects along major technical innovations were adopted by southern Levant rural communities during the Late Chalcolithic period (4500–3900 cal. BC), also called the Ghassulian (Gilead 2011). Two extreme scenarios were elaborated, respectively, by Gilead (1988) and Levy

---

[2] Ties can be strong or weak. In analytical sociology, "strong ties describe frequently activated relationships (such as family/kin ties) whereas weak ties are used to describe infrequently accessed connections (acquaintances)" (Collar et al. 2015, p. 23).

(1995) for explaining both the presence of prestige objects and the absence of any influential central site (Rowan and Golden 2009). According to Gilead (1988), the Ghassulian societies were egalitarian given the absence of monuments or building indicating chiefs, tombs with rich funerary objects, hierarchy between and within the villages, and common storage facilities. The new objects in copper, basalt, or exotic stones may testify to ritual activities without necessary elites responsible for their production. Against this hypothesis, Levy (Levy and Holl 1988; Levy 1995) considers that the material culture of the Ghassulian societies points to ranked social hierarchies with a politico-religious power, controlling the resources of the territories and redistributing them. Corresponding evidence includes specialized craftsmanship, burial caves with rich goods, pilgrimage places, hierarchized organization of settlements, and, finally, the function of metal objects pointing toward elites. Another hypothesis, based on the ceremonial function of wheel-made bowls, suggests that the invention of the potter's wheel emerged following the demand of wheel-shaped bowls by an elite (Roux 2010). In this hypothesis, the term "elite" has remained vague and does not imply any specific politico-religious system. The debate is still vivid as more evidence of so-called prestige objects points toward complex networks of production and distribution and interconnected ties between communities sharing similar norms (Rosenberg et al. 2016; Rowan and Golden 2009). More generally, the debate relates to both the connections between the Ghassulian communities and the historical process which led to major changes in the material culture.

This paper is organized in three sections. The first one discusses similarity variables for assessing social relationships between sites. The second section analyzes archeological data and shows how technological attributes are meaningful qualitative data for establishing social links between sites and revealing social topology, i.e., the overall arrangement of social ties in which actors are embedded, as well as population structure, i.e., instances where individual subpopulations/groups exhibit low within and high between variability (Shennan et al. 2015). The third section discusses modeling evolution processes on the basis of qualitative variables and explaining these processes by reference to sociological models.

Technological Traditions: Similarity Attributes and Social Connections

As noted before, the variables used in archeological models are mainly morphometric and stylistic variables (i.e., size and shape of objects, decorating elements). These are used to create links between sites and assess against temporal and spatial data whether these links indicate either a cultural group or connections at a macroregional scale.

In network analysis, social relationships between sites are mostly inferred from the co-occurrence of artifacts (the more shared artifacts, the stronger the relationships;

Collar et al. 2015). The underlying principle is that shared similar artifacts express interactions between sites or within sites and, therefore, social relationships (Coward 2013). In other words, similar artifacts are variables to measure social interactions, while the resultant network expresses a network of exchanges. Indeed, in network studies interactions are mainly considered as the expression of exchange-based relationships (Gjesfjeld and Phillips 2013). Now, this link between interactions and exchange is not straightforward. Indeed, material culture presents different aspects, and the presence of a same type of object on two sites can be due to numerous reasons: one of the two objects may be imported, copied, and made by migrants, on brokers' requests (indirect contacts), by individuals of a social group scattered over a large region, etc. Depending on the situation, similarity objects express different types of interactions between groups: exchange-based relationships, market distribution, ethnic affiliation, matrimonial alliances, movements of individuals or populations, etc. It means that when listing similarity attributes, if they are of one kind (one type of pots), which is the case most of the time (Östborn and Gerding 2014, p. 79; Östborn and Gerding 2015), the networks obtained express relationships which can be of different types.

In this sense, the issue relates to both the type of social interactions one wants to measure and the variables to use depending on the type of social interactions to highlight. As said before, much of the archeological use of network analysis targets interregional connections between sites and exchange-based relationships. On the contrary local networks have been less considered (Knappett 2018). When studied, these local networks are modeled based on probable interactions among sites sharing consumption of objects, for example, given similarities in proportion to decorated ceramics present at pairs of sites (Borck et al. 2015; Mills et al. 2013). Interactions are thus measured through the sharing of attributes, but the social content of these interactions is not measured because similarity objects can be the result of different types of interactions, making difficult to explore the strength of the links in between and within the sites as well as network topologies.

Variables to Socially Linked Individuals/Groups

In contrast, we propose to consider the ceramic *chaîne opératoire* as a robust variable for assessing degrees of connections between sites. It is defined as the series of actions that transform raw material into finished product, either consumption object or tool (Creswell 1976, p. 13). Its description implies the characterization of objects in terms of manufacturing methods, techniques, and tools. A method is defined as an ordered sequence of functional operations carried out by a set of elementary movements for which different techniques can be used. A technique is defined as the physical modalities used to transform the raw material into a finished product. Techniques are in limited number, contrary to methods whose variability is – theoretically – infinite. Thus if techniques can be the object of convergence, on the contrary, methods are more likely to be specific. In other words, this is the unique combination of sequences, gestures, and techniques that makes technological traditions highly cultural and unique to social groups, therefore distinguishing between traditions linked through the transmission of information and convergent solutions to specific situations (Shennan 2002, p. 73).

The transmission of *chaînes opératoires* involves constraints. Indeed, studies in the anthropology of techniques and cultural transmission show that transmission of craft techniques necessarily requires social learning, that is to say, learning from tutors contrary to individual learning (O'Brien and Bentley 2011, 317). Guided transmission of skills consists in educating the learner about the information available in the environment, be it the properties of the material, the tools used, or the effects of the gestures employed (Bril 2002). This guidance not only facilitates the learning process but also participates in the reproduction of the task (Tehrani and Riede 2008). It is the key to the cultural transmission of ways of doing things. The tutor is usually selected within one's social group (Gosselain 2000; Shennan 2013; Shennan and Steele 1999). As a result, technological traditions signal that individuals having the same tradition belonged to the same once "community of practice," i.e., a community sharing ways of doing (Lave and Wenger 1991). This term might seem awkward when used for connecting sites since communities of practice are defined as groups of people who interact regularly (Wenger 2000), whereas in archeology it is problematic to demonstrate regularity of interactions. "Community of practice" is to be better understood as a process, a mechanism which explains how traditions are created (in the course of learning), perpetuated, or modified (Gosselain 2008). In archeology, similarity between ways of doing can be seen as the result of this process, however spread out in time and space, and therefore signaling before all communities made up of individuals who learned and taught a same craft tradition within the framework of historically determined social links. Spatial patterns of these communities can be the result of both historical and sociocultural processes: population expansion and/or sociocultural circulation of individuals (e.g., through matrimonial alliances). On the contrary, dissimilarity in craft techniques between sites signals different communities, that is, communities whose individuals do not share the same practices and therefore are not part of the same social group. Similarity or dissimilarity in craft techniques can thus link sites and bring to light social communities and locally driven networks, similarity indicating strong ties, and dissimilarity, weak ties. The overall spatial arrangement of technological traditions reflects population structure (Hodder 1985; Roux et al. 2017; Stark et al. 2008; Stark 1998).

Let us note that in ceramic technological traditions, the longest stage to learn is the forming stage because of the general difficulty of mastering motor skills (Bril 2002; Ericson and Lehman 1996). Forming techniques are taught with a tutor over years usually within private spaces, while shapes, decorative features, or even clay recipes can be learned through individual learning after seeing objects in public spaces and/or discussing with retailers (e.g., interactions with shopkeepers) (Roux 2015). As a consequence, forming techniques tend to be more resistant to change than easily transmissible traits such as style (shapes and decor of objects) (Gallay 2007; Gelbert 2003; Gosselain 2000; Hegmon 1998; Mayor 2010; Roux 2015;

Stark et al. 2000). In this respect forming technique is a better variable to connect over time individuals/communities from the same social group than shapes and decoration whose evolutionary mechanisms make them more subject to rapid changes and diversity, even within the same social group. It must therefore be remembered that, when submitted to evolution (either through phylogenesis or ethnogenesis process), the different stages of the *chaînes opératoires* are meant to change at different rhythms because they are subject to different mechanisms of change. Each of them should be considered as a distinct variable, signaling different types of interactions depending on their co-occurrence.

Variables to Measure Network Topology

In social network analysis, assessing connections between groups implies not only to assess the similarity between groups but also to examine the embeddedness of the network and therefore the network topology. Embeddedness is "an indicator of how a particular individual or social group will socially interact by either choosing to network with many other individuals or only a few" (Borck et al. 2015, p. 37). The examination of embeddedness quantifies "how a particular group is likely to interact with its neighbors at a given point in time and how that may affect the network and actors during later temporal intervals" (Borck et al. 2015, p. 37). Different quantifying methods are used. For example, ceramic types are apportioned into time intervals (Mills et al. 2013). Measures of similarity between sites are based on the relative percentages of apportioned ceramic types between pairs of sites. In order to calculate connections inside the regions and outside the regions, and therefore embeddedness at the population level, groups are made based on independent criteria (archeological and geographical boundaries). These quantified measures require a high chronological resolution which unfortunately is not often the case.

A qualitative approach to embeddedness is the composition of the ceramic assemblages at the macro-regional scale (Roux 2016, chap. 4). This composition may testify to interactions between communities at different scales. In this aim, ceramic assemblages are analyzed in terms of techno-petrographic homogeneity *versus* heterogeneity. Theoretically, we distinguish two categories of ceramic assemblages – homogeneous and heterogeneous assemblages – on the basis of techno-petrographic groups (groups including ceramics made the same way with the same clay material). Simple homogeneous assemblages (made of one techno-petrographic group) characterize sites whose producers belong to a homogeneous social group and use clay on site. Complex homogeneous assemblages (*n* techno-petrographic groups) characterize sites with distinct groups of producers using different clay sources close from the site. Simple heterogeneous assemblages reveal diverse techno-petrographic groups. Petrographic heterogeneity indicates ceramic production on a meso-regional scale, and the low variability of this heterogeneity allows us to define the region of the clay material sources. Complex heterogeneous assemblages reveal diverse techno-petrographic groups. Due to petrographic heterogeneity, and marked variability, it is not possible to define a single region, and ceramic production sites are dispersed over a macro-regional scale. Simple and complex heterogeneous assemblages testify at a given point in time to the presence of consumers originating from the meso- or macro-region of the site. The former point to embeddedness at the regional level and the latter to embeddedness at the population level.

In brief, the techno-petrographic analysis of ceramic assemblages at a macro-regional scale should enable us to highlight whether there are movements of individuals between sites and whether these movements indicate interactions. In a macro-region where sites are recognized as epicenters of interactions, these sites indicate strong network embeddedness.

Connecting Ghassulian Sites with Technological Traditions

In this section, we show how the *chaîne opératoire* approach enables us to link the Ghassulian sites of the southern Levant. The artifacts characterizing the Ghassulian culture as a "coherent culture" and connecting the sites are first recalled, followed by the technological ceramic analysis.

Shared Ceremonial Objects in the Southern Levant During the Ghassulian Period

The Ghassulian sites of the southern Levant (Fig. 1) were occupied by agropastoral communities who indisputably shared mundane material culture (repertoire of ceramic vessels, lithic tools, grinding material), iconographic motifs, mortuary practices, and ceremonial objects, suggesting a coherent culture (Lovell and Rowan 2011; Rowan and Golden 2009).

The size of the sites and their spatial patterning do not point toward hierarchy or some sort of centrality. A few sites range up to 10 ha; a few others are large (c. 4–5 ha). However, most of the sites are small (c. <1–2 ha). The relationship between larger and smaller sites remains unknown (Levy et al. 2006). The village houses are broadly similar from the Golan to the Negev. Within sites, there is no size differentiation, hierarchy in room sizes, or obvious elite areas.

The social relationships between sites are difficult to assess mostly because of a lack of chronological control. Indeed, the Ghassulian period of the southern Levant is probably divided in two phases, as shown by researches in the northern Negev (Gilead 2011). However, these phases have not been distinguished on most sites because there is practically no change in stratigraphy and the artifactual assemblages (Gilead 1994; Rowan and Golden 2009); at this stage, one can just keep in mind that variability in some patterns could be chronological.

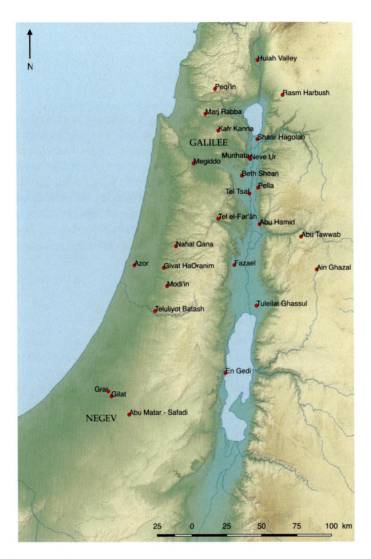

Fig. 1 Sites and regions of the southern Levant cited in the text

This being said, there are shared objects between sites, providing evidence for local networks. The most conspicuous are the so-called ceremonial/prestige objects whose distribution highlights patterns at different regional scales. They are the ones considered here as they testify to shared symbolic norms among which some emerged along with major technical innovations, such as wheel-shaped bowls and the wheel-coiling technique and copper objects and the lost-wax technique. They include basalt vessels, wheel-shaped bowls, perforated flint tools, stone violin figurines, maceheads, copper objects, and ivory objects. As we shall see, basalt bowls, wheel-shaped bowls, violin figurines, and maceheads connect sites from all over the southern Levant, while perforated flint tools and ivory and copper objects are more regionally distributed.

Regional Distribution

Perforated flint objects are unique items whose distribution is marked geographically. They are interpreted as prestige objects because they require high knapping skills, show surprisingly limited use, and are found in both domestic (settlement) and ritual contexts (burial caves) (Rosen 1997; Rosenberg and Shimelmitz 2017). Their distribution reflects a northern network, including sites of the Jordan Valley.

Ivory and copper items are the two other categories of prestige/ceremonial objects to circulate within a geographically constrained network. They are found mainly in the southern regions. Copper artifacts, probably produced in ritualized contexts, were made locally as shown by production artifacts on the Beersheva sites (Goren 2014; Gilead and Gošić 2015).

Macro-regional Distribution

However, there is a distribution of ivory and copper items at the macro-regional scale, although it is restricted to three pivotal collective burial caves: Nahal Qanah, Giv'at Ha-Oranim, and Peqi'in. On the basis of their context of finding, metal ceremonial objects have been interpreted as a burial kit of graves (along goods like ivory, basalt, textiles, and ceremonial ceramics) involving multiple burials and, oftentimes, secondary burials associated with ossuaries (Golden 2009). In this respect, they indicate both sharing of mortuary practices and norms over a wide area, from Negev up to Galilee, even though their fabrication was restricted to the northern Negev.

Basalt vessels are found in most Ghassulian settlements throughout the southern Levant, most typically in domestic contexts. They are supposed to be prestige/ceremonial objects, given the skills involved in their manufacturing and the long distance between the basalt sources (east in Jordan or north in the Galilee and Golan) and many of the sites where they have been found (Chasan and Rosenberg 2018). Moreover, numerous basalt vessels are reported in eight Ghassulian burial caves arguing in favor of their interpretation as highly valued objects (Chasan and Rosenberg 2018). Their wide dispersion throughout the region, in ceremonial or domestic sites, reflects a far-reaching macro-regional network (Rosenberg et al. 2016).

Wheel-shaped bowls are also found all over the southern Levant in settlements and funerary contexts. They are supposed to be ceremonial objects (lamps as shown by soot traces). The main arguments are as follows: (a) the wheel-coiling technique implies specialized skills that are in play only for this category of bowls, henceforth making them highly valued objects, and (b) their presence in funerary contexts is systematic whatever the type of burial and the geographical zone (Roux 2003).

These bowls were produced on-site. They are supposed to have been made by itinerant potters (Roux and Courty 2005). In this regard, they testify to movements of individuals from site to site arguing in favor of strong connections between the sites.

The violin figurines and the stone maceheads, whose raw material is varied and originates possibly from different geographical zones, have been found in a limited number of sites but throughout southern Levant, from north to south. Like the basalt vessels, they testify to shared norms through the consumption of the same category of objects by geographically dispersed communities.

Distribution of Ceremonial Objects at Different Scales

In summary, specific patterns of distribution of ceremonial objects reflect networks at different scales, regional *versus* macro-regional networks, and different time periods.

Regional networks distinguish roughly between northern/eastern and southern/western clusters of sites. The northern/eastern cluster gathers sites in northern Israel (Golan, Hula valley, the Galilee), the Middle Jordan Valley, northern Jordan, and southern Syria, through the circulation of perforated flint tools. The southern/western cluster gathers sites mainly from the Shephelah – littoral coast southward – through the manufacture and use of copper objects and the circulation of ivory objects. Within each of these regions, smaller-scale geographical zones can be highlighted when examining the provenance of the ceramics found on ceremonial or burial places. Thus petrographic analysis of Gilat, a supposedly ceremonial place in the northern Negev, highlights that it was a center for the northern Negev and the Judean mountains (Goren 1995), whereas En Gedi, a shrine above the Dead Sea, was frequented by individuals from the Judean mountains only (Goren 1995; Roux and Courty 2007). In this respect, regional ceremonial places could correspond to local cults coexisting with a macro-regional cult (Gilead 2002). The unique basalt pillar figurines of Golan, found in household contexts, echo these local cults; similarly, regional burial caves such as in Azor where ceramic bowls have been found to be produced within a meso-region (30 km around) echo local networks (Roux and Courty 2007).

The macro-regional network testifies to the sharing of the same ceremonial norms through the use of basalt bowls, wheel-shaped bowls, violin figurines, and maceheads. This macro-regional network is apparently active during the whole Ghassulian period (early and late). It is also visible in the funerary goods found at three pivotal burial caves: Peqi'in, Nahal Qanah Cave, and Giv'at Ha-Oranim. These caves (probably late Ghassulian), characterized by multiple burials deposited over time, are exceptional in the way they gather ceremonial objects that are either shared at the macro-regional scale (basalt bowls, wheel-shaped bowls) or at the regional scale only (perforated flint object from the northern network, copper and ivory items from the southern network) (Rosenberg and Shimelmitz 2017). They thus indicate shared ceremonial norms between communities integrated also within distinct regional networks, notwithstanding a variety of burial practices within southern Levant.

In order to debate whether the sharing of norms as expressed by ceremonial artifacts across the southern Levant was encouraged by a hierarchy or by the social structure itself, local connections between sites remain to be drawn. These connections will be tentatively highlighted on the basis of the ceramic *chaînes opératoires* used for making the mundane ceramic vessels.

Ghassulian Local Networks: The Ceramic Chaînes Opératoires

For assessing qualitatively the degree of interactions between sites and drawing hypothesis about social topology, the *chaînes opératoires* carried out for making the mundane containers constitutive of the ceramic assemblages have been analyzed in terms of similarity/dissimilarity.

The sites are located in different parts of the southern Levant[3] (Fig. 1). Ceramic assemblages belong to well-established Ghassulian horizons even though the chronological span can cover a few hundred years.

The *Chaînes Opératoires*: Connecting Sites

Results highlight that the same *chaîne opératoire* was carried out by all the Ghassulian communities of the southern Levant (Roux 2019). It entails the following operations. Clay material is usually mixed with 20–30% coarse mineral grains whose size depends on the thickness of the walls. Petrographic studies highlight the local production at almost all sites studied (Rowan and Golden 2009). The bases are modeled into a disc shape from a lump of clay, the edges are raised, and an inner peripheral coil is placed against the edges on the disc. The next successive coils are fixed by apposition against the inner face. Once the body is shaped, the rim is thinned and shaped with a wet piece of cloth. After shaping the rim, the inner face is smoothed with a dry or wet soft tool. The vessel is left dried until leather-hard consistency. Elements are applied at that point: decorative bands, handles, and, for all the vessels, an extra peripheral coil around the external base, probably as a reinforcement piece. A coating is then applied on the outer face for the closed vessels and on both sides for the open vessels. The decoration of the bands by finger impression is made after the coating, as well as the perforation of the handles.

There are regional variants to this *chaîne opératoire*, like, in the north, the application of a red slip on the coating. These variants still need to be recorded more

---

[3] The studied sites are in the Jordan Valley and the Dead Sea basin (Teleilat Ghassul, Fazael, Abu Hamid, Pella, Tel el-Far'âh [N., cave U], Neve Ur), in the Negev (Abu Matar, Safadi, Grar), in the Shephelah (Modi'in), in the coastal plain (Azor), in the Galilee (Kafr Kanna, Levels 112–115; Megiddo, stratum 5), in the Hula valley (Tel Teo, Turmus), and in the Golan (Rasm Harbush).

systematically and should enable us to cluster groups interacting preferentially with one another.

The *Chaînes Opératoires*: Population Structure

The Ghassulian *chaîne opératoire* contrasts with the *chaînes opératoires* used during the fifth to fourth millennium BC by the neighboring populations. In Egypt, the clay paste is tempered with animal dung, the bases are made from spiraled coils, the bodies are made with horizontally superimposed coils, and the external faces are burnished. To the north of the southern Levant, several traditions are used, including shaping by modeling or by adding large coils. None of them use clay coating (Baldi 2017).

These different traditions do not stem from temporal and/or spatial factors: there are no similar technical elements which could signal a common origin. In this respect, the technological analysis of the ceramics highlights a population structure distinguishing the southern Levant population from its neighbors who were social groups within which other ways of doing ceramics were transmitted.

Composition of the Ceramic Assemblages: Social Topology

The issue lies in assessing whether the Ghassulian communities, distributed over a wide regional area, sharing the same ceramic tradition as well as ceremonial objects, but made of distinct groups based on the geographical distribution of some techno-stylistic variants (shapes, slip), were interacting at the population scale within an embedded social network.

A techno-petrographic study of ceramic assemblages belonging to sites distributed all over the southern Levant shows that there are three main categories of assemblages: (a) homogeneous assemblages testifying to interactions at the scale of the village, (b) simple heterogeneous assemblages testifying to interactions at the regional scale, and (c) complex heterogeneous assemblages testifying to interactions at the population scale (Roux and Courty 2007).

Simple heterogeneous assemblages indicating interactions at the regional scale are met in shrines or burial sites (e.g., Gilat, En Gedi, Azor) (Goren 1995). Complex heterogeneous assemblage is met only at one site, Abu Hamid in the Middle Jordan Valley. The techno-petrographic analysis of the ceramic assemblage shows that all the recipients come from all over the southern Levant (Roux and Courty 2007). In the Late Chalcolithic cultural context, Abu Hamid has been interpreted as a gathering/pilgrimage site, that is to say, a place frequented by people who come from all over the southern Levant. In this respect, the ceramic assemblage of Abu Hamid suggests that Ghassulian sites were connected at the population level at a given point in time and therefore that the Ghassulian population was a homogeneously mixed population (each individual can interact with one another) within a tight embedded social network.

Modeling and Explaining Evolution Processes: Phylogenetic and Sociological Models

In the previous section, we have seen that qualitative variables such as the *chaînes opératoires* allow to socially connect the Ghassulian sites of the southern Levant and unveil the social topology as well as the population structure. In a general similarity network analysis, i.e., "a flexible framework in which all kinds of similarity relations (including well defined differences) can be used as proxies for causal or social relationships and define links between archaeological contexts" (Östborn and Gerding 2014, p. 87, 2014), the Ghassulian *chaîne opératoire* can thus be added as an attribute whose presence or absence, combined with types of vessels, will be a major indicator of the relational structure of the society.

Modeling evolution processes is another issue discussed below. We distinguish between (1) modeling evolution processes through phylogenetic models for testing, in our case study, how the Ghassulian homogeneous mixing population was created and (2) explaining historical changes by transferring regularities highlighted by sociological models on archeological data.

Phylogenetic Models and Technology

Technological traditions can also be considered as relevant variables to establish cultural lineages, defined as "traditions linked by historical continuity based on the transmission of information through time" (Shennan 2002, p. 72). Indeed, traits describing the *chaînes opératoires* are particularly relevant as opposed to shapes because, as previously said, they are traits transmitted between producers from the same social group and therefore quite stable, contrary to shapes that are likely to evolve with the evolution of the demand (needs), to be easily copied (without necessary social learning) and consequently to change rapidly. Identifying cultural lineages implies the study of assemblages of different periods, preferentially from the same sites, and a detailed analysis of the *chaînes opératoires* in order to trace socially learned ways of doing vessels handed down over the centuries.

The characterization of cultural lineages is a qualitative approach but can be coded with computational tools for quantitative output. In this case, cladistics is the preferred approach, as it allows for the measurement of the phylogenetic signal linking assemblages (O'Brien et al. 2001, 2003). The construction of a phylogenetic tree is based on a simple fundamental principle: lineage with modification. The application of the phylogenetic tree to techniques allows for the modeling of the diversity of *chaînes opératoires*, their evolution and their kinship in a cultural group (phylogenesis), and any possible extra-cultural transfers (ethnogenesis) (Manem 2008).

In the case of the Ghassulian societies of the southern Levant, the recognition of a large rural social group occupying the whole southern Levant during the Ghassulian period raises the issue of how a wide social network sharing a same tradition developed. The hypothesis is that the macro-regional network finds its roots in an ancestral social network, whereas the regional networks may have emerged over time through geographical connectivity only.

To work out this hypothesis, ideally, phylogenetic modeling could measure the evolution of the ceramic traditions over centuries. A preliminary qualitative study of the ceramic assemblages from a few sites[4] dated from the end of the seventh to early fifth millennium BC suggests the existence of phylogenetic links between the Neolithic and the Chalcolithic assemblages. The main *chaîne opératoire* involved in the making of the mundane vessels is the same as the one used during the Ghassulian period: the clay paste is prepared with a high proportion of coarse mineral temper, the bases are modeled in the form of a disc and edges raised, small coils are laid inward, reinforcing coils are laid around the outer bases, and, finally, walls are clay coated. This *chaîne opératoire* is the stable element while other traits evolved over time (in particular shapes and decor). It represents the ancestral trait linking the Neolithic and Ghassulian ceramic assemblages testifying to the transmission of a same way of doing from generation to generation over more than two millennia and therefore to the relationship by descent between the related communities. This relationship by descent amounts to saying that the ancestors of the Ghassulian groups were the Neolithic communities of the southern Levant. If this hypothesis is correct, ties of kinship, to varying degrees, may have united all the individuals forming the Ghassulian society of the southern Levant, explaining the Ghassulian embedded social network.

The evolving traits, creating variants between regions over time, are the derived traits. They may testify to the spatial spreading and splitting of the social groups over the two millennia separating the seventh and fifth millennia populations. Phylogenetic modeling should help to test this hypothesis by highlighting the derived traits and their possible relationships with the emergence of regionally centered ceremonial/burial places. Among the derived traits, not only technical variants but also degrees of similarities and differences in ceramic shapes between sites as a function of interactions should be taken into account.

Sociological Models for Explaining Evolution Processes

Phylogenetic models describe modification of material culture in terms of transmission. The question is how to explain these modifications in sociological terms. We believe that simulations of archeological data are not meant to provide explanations to correlations between social structure and changes. Rather, explanations have to

---

[4]They include sites from the Jordan Valley (Abu Hamid, Tel Tsaf, Beth Shean XVIII, Munhata, Shaar Hagolan), the Shephelah (Teluliyot Batash), and the Jordanian plateau (Ain Ghazal, Abu Thawwab).

be looked for in analytical sociology whose models provide explanatory founded regularities transferable by analogy to archeological data as we shall see below.

Let us return to our case study and summarize the archeological data of the Ghassulian southern Levant and their sociological interpretation: new ceremonial norms, visible in the making of new objects with new techniques, appeared and were shared at the scale of the southern Levant, whereas there is limited evidence for hierarchical formation and centralized political power (Rowan and Golden 2009). This wide sharing of new norms raises the question of the historical process by which the Ghassulian population rapidly evolved shared social conventions. The technological analysis of the Ghassulian ceramic assemblages reveals that a same *chaîne opératoire*, from the clay preparation to the firing, was shared at the scale of the southern Levant. In this respect, the widely shared ceramic tradition of the southern Levant testifies to a same social group wherein it has been transmitted. Technological boundaries have been maintained over centuries suggesting the maintenance of social boundaries, possibly through endogamous matrimonial alliances. The site of Abu Hamid where the different communities met at a given point in time suggests that this broad social group was a homogeneous mixing population (each individual can interact with one another).

If these hypotheses are correct, we are then in a position to use the results obtained by sociologists who have recently worked on the social topology required for populations to share new norms (Centola and Baronchelli 2015). More specifically, they questioned the social structures favorable to the adoption of new social conventions at the population level without large-scale coordination. Experiments were made on the web with players who were presented with a picture of someone and who had to agree on the name to give. The hypotheses were that repeated interaction produces collective agreement among a pair of players and, following a broad range of formal approaches, that "the connectivity of the actor's social networks can influence the collective dynamics of convention formation, ranging from the emergence of competing regional norms that inhibit global coordination to the rapid growth of universally shared social conventions" (Centola and Baronchelli 2015, p. 1990). Three types of networks were tested: (1) spatially embedded social topologies (interactions between actors close spatially), (2) randomly connected topologies (random interactions), and (3) homogeneously mixing populations (each individual can interact with one another). The results show that the network structure that promotes the emergence of shared social norms is the one with the higher connectivity between individuals, i.e., a homogeneously mixing population made up of individuals able to interact with all the individuals of the community. In other words, results show that shared conventions can emerge in complex decentralized systems, without coordinated leadership, when there is a homogenously mixing population and therefore multiple interactions within a dense network structure.

By analogy, it is possible to propose that the shared Ghassulian ceremonial norms emerged as the result of network connectivity, without any large-scale coordination. The rationale is that if archeological data testify to a homogeneous mixing population, then by analogy with the sociological model, it is possible to interpret the emergence of common norms as the result of intense interactions, without any global-political coordination. This hypothesis is not contradicted by the ceremonial objects themselves: the major technological innovations (the lost-wax technique and the wheel-coiling technique) were exclusive to ceremonial objects; therefore, they point to belief depositories (e.g., shamans) as suggested by Gilead (2002), but not to any specific politico-religious power. The validation of the archeological interpretation can be evaluated against the validity of the regularity (the model) tested by the study of its generating mechanisms. Indeed, archeological interpretation proceeds by analogy, and in the case of explaining historical processes, the data are too lacunar to test through simulations all the variables which may have played a role in the historical process itself. Only in present-day situations, it is possible to combine empirical studies and simulations for testing the causal role of different variables and interactions.

Conclusion

In this paper, I have questioned qualitative variables and their use into computational models for revealing the relational structure of societies, knowing that social structures represent conditions of actualization of evolutionary phenomena. I argue that not all the similarity attributes can socially connect sites; among them, technological traditions are the best candidates. They allow not only to trace social connections between sites but also to characterize social topology and population structure. Moreover, they are powerful variables to establish phylogenetic links. They can easily integrate computational models, knowing that any qualitative data can be subsequently quantified (even narratives, see Manzo et al. 2018). The validation of the models obtained lies in the well-founded social significance of the variables and the qualitative analysis of the archeological material.

Validation of the hypotheses explaining evolution processes depends not only on the variables taken to model archeological data but also on the founding of the referential model. Indeed, archeological data are too lacunar and polysemic for obtaining explanations that we could empirically validate. The way out is to interpret them with the help of reference regularities obtained in the domain of present-day societies through simulation methods whose results can be validated against empirical data. The validation of the archeological interpretation lies in both the analogical operation and the founding of the reference regularity transferred to archeological data.

To conclude, the power of computational models for interpreting evolution processes needs not anymore to be demonstrated. However, these models require meaningful variables, and, for this purpose, qualitative analyses of archeological material are more than necessary. Unfortunately, these analyses remain too few, and proper integration of combined relevant proxies is still pending.

References

Axelrod, R. (1997). The dissemination of culture a model with local convergence and global polarization. *Journal of Conflict Resolution, 41*(2), 203–226.

Baldi, J. S. (2017). Collections céramiques du Musée de Préhistoire Libanaise : une étude technique. ArchéOrient – Le Blog. http://archeorient.hypotheses.org/7431.

Bentley, R. A., & Shennan, S. J. (2003). Cultural transmission and stochastic network growth. *American Antiquity, 68*(3), 459–485.

Bettinger, R. L., & Eerkens, J. (1999). Point typologies, cultural transmission, and the spread of bow-and-arrow technology in the prehistoric great Basin. *American Antiquity, 64*(2), 231–242.

Blake, E. (2014). Dyads and triads in community detection: A view from the Italian bronze age. *Les nouvelles de l'archéologie, 135*, 28–32.

Borck, L., Mills, B. J., Peeples, M. A., & Clark, J. J. (2015). Are social networks survival networks? An example from the late pre-Hispanic US Southwest. *Journal of Archaeological Method and Theory, 22*(1), 33–57.

Bril, B. (2002). L'apprentissage de gestes techniques: ordre de contraintes et variations culturelles. In B. Bril & V. Roux (Eds.), *Le geste technique. Réflexions méthodologiques et anthropologiques* (pp. 113–150). Ramonville Saint-Agne: Editions érès.

Brughmans, T. (2010). Connecting the dots: Towards archaeological network analysis. *Oxford Journal of Archaeology, 29*(3), 277–303.

Brughmans, T. (2013). Thinking through networks: A review of formal network methods in archaeology. *Journal of Archaeological Method and Theory, 20*(4), 623–662.

Cegielski, W. H., & Rogers, J. D. (2016). Rethinking the role of agent-based modeling in archaeology. *Journal of Anthropological Archaeology, 41*(Supplement C), 283–298.

Centola, D. (2015). The social origins of networks and diffusion. *American Journal of Sociology, 120*(5), 1295–1338.

Centola, D., & Baronchelli, A. (2015). The spontaneous emergence of conventions: An experimental study of cultural evolution. *Proceedings of the National Academy of Sciences, 112*(7), 1989–1994.

Centola, D., & Macy, M. (2007). Complex contagions and the weakness of long ties1. *American Journal of Sociology, 113*(3), 702–734.

Charbonneau, M. (2018). Technical constraints on technological evolution. In M. J. O'Brien, B. Buchanan, & M. I. Eren (Eds.), *Convergent evolution and stone-tool technology* (p.73-89). Cambridge: MIT Press.

Chasan, R., & Rosenberg, D. (2018). Basalt vessels in Chalcolithic burial caves: Variations in prestige burial offerings during the Chalcolithic period of the southern Levant and their social significance. *Quaternary International, 464*, 226–240.

Collar, A., Coward, F., Brughmans, T., & Mills, B. J. (2015). Networks in archaeology: Phenomena, abstraction, representation. *Journal of Archaeological Method and Theory, 22*(1), 1–32.

Coward, F. (2013). Grounding the net: Social networks, material culture and geography in the Epipalaeolithic and Early Neolithic of the Near East (~ 21,000–6,000 cal BCE). In C. Knappett (Ed.), *Network analysis in archaeology: New regional approaches to interaction* (pp. 247–280). Oxford: OUP.

Creswell, R. (1976). Techniques et culture, les bases d'un programme de travail. *Techniques & Culture, 1*, 7–59.

Dunnell, R. C. (1978). Style and function: A fundamental dichotomy. *American Antiquity, 43*, 192–202.

Eerkens, J. W., & Lipo, C. P. (2005). Cultural transmission, copying errors, and the generation of variation in material culture and the archaeological record. *Journal of Anthropological Archaeology, 24*(4), 316–334.

Eerkens, J. W., & Lipo, C. P. (2007). Cultural transmission theory and the archaeological record: Providing context to understanding variation and temporal changes in material culture. *Journal of Archaeological Research, 15*(3), 239–274.

Ericson, K. A., & Lehman, A. C. (1996). Expert and exceptional performance: Evidence from maximal adaptation to task constraints. *Annual Review of Psychology, 47*, 273–305.

Flache, A., & Macy, M. W. (2011). Small worlds and cultural polarization. *The Journal of Mathematical Sociology, 35*(1–3), 146–176.

Flache, A. (2018). Between monoculture and cultural polarization. Agent-based models of the interplay of social influence and cultural diversity. *Journal of Archaeological Method and Theory, 25*(4), 996–1023.

Gallay, A. (2007). The decorated marriage jars of the inner delta of the Niger (Mali): Essay of archaeological demarcation of an ethnic territory. *The Arkeotek Journal (www.thearkeotekjournal.org), 1*(1).

Gallay, A. (2011). *Pour une ethnoarchéologie théorique*. Paris: Editions Errance.

Gandon, E., Roux, V., & Coyle, T. (2014). Copying errors of potters from three cultures: Predictable directions for a so-called random phenomenon. *Journal of Anthropological Archaeology, 33*, 99–107.

Gelbert, A. (2003). *Traditions céramiques et emprunts techniques dans la vallée du fleuve Sénégal. Ceramic traditions and technical borrowings in the Senegal River Valley*. Paris: Editions de la Maison des sciences de l'homme, Editions Epistèmes.

Gilead, I. (1988). The Chalcolithic period in the Levant. *Journal of World Prehistory, 2*, 397–443.

Gilead, I. (1994). The history of the Chalcolithic settlement in the Nahal Beer Sheva area: The radiocarbon aspect. *Bulletin of the American Schools of Oriental Research, 296*, 1–13.

Gilead, I. (2002). Religio-magic behavior in the Chalcolithic period of Palestine. In S. Ahituv & E. D. Oren (Eds.), *Aharon Kempinski memorial volume: Studies in archaeology and related disciplines*. Beersheva: Ben-Gurion University of the Negev Press.

Gilead, I. (2011). Chalcolithic culture history: The Ghassulian and other entities in the southern Levant. In J. L. Lovell & Y. Rowan (Eds.), *Culture, chronology and the Chalcolithic. Theory and transition* (pp. 12–24). Oxford and Oakville: CBRL and Oxbow Books.

Gilead, I., & Gošić, M. (2015). Unveiling hidden rituals: Ghassulian metallurgy of the Southern Levant in light of the ethnographical record. In K. Rosinska-Balik, A. Ochal-Czarnowicz, M. Czarnowicz, & J. Debowska-Ludwin (Eds.), *Copper and trade in the South-Eastern Mediterranean. Trade routes in the Near East in antiquity* (pp. 25–38). Oxford: Archaeopress.

Gjesfjeld, E., & Phillips, S. C. (2013). Evaluating adaptive network strategies with geochemical sourcing data: A case study from the Kuril Islands. In *Network analysis in archaeology: New approaches to regional interaction* (pp. 281–305).

Golden, J. (2009). New light on the development of Chalcolithic metal technology in the southern Levant. *Journal of World Prehistory, 22*(3), 283–300.

Goren, Y. (1995). Shrines and ceramics in chalcolithic Israel: The view through the petrographic microscope. *Archaeometry, 37*(2), 287–305.

Goren, Y. (2014). Gods, caves, and scholars: Chalcolithic cult and metallurgy in the Judean Desert. *Near Eastern Archaeology (NEA), 77*(4), 260–266.

Gosselain, O. (2000). Materializing identities: An African perspective. *Journal of Archaeological Method and Theory, 7*(3), 187–217.

Gosselain, O. (2008). Mother Bella was not a Bella. Inherited and transformed traditions in Southwestern Niger. In M. Stark, B. Bower, & L. Horne (Eds.), *Cultural transmission and material culture. Breaking down boundaries* (pp. 150–177). Tucson: Arizona University Press.

Granovetter, M. (1983). The strength of weak ties: A network theory revisited. *Sociological Theory, 1*(1), 201–233.

Hamilton, M. J., & Buchanan, B. (2009). The accumulation of stochastic copying errors causes drift in culturally transmitted technologies: Quantifying Clovis evolutionary dynamics. *Journal of Anthropological Archaeology, 28*(1), 55–69.

Hegmon, M. (1998). Technology, style, and social practice: Archaeological approaches. In M. T. Stark (Ed.), *The Archaeology of social boundaries* (pp. 264–279). Washington, DC: Smithsonian University Press.

Hodder, I. (1985). Boundaries as strategies: An ethnoarchaeological study. In *The archaeology of frontiers and boundaries*. New York: Academic Press.

Jordan, P., & Shennan, S. (2003). Cultural transmission, language, and basketry traditions amongst the California Indians. *Journal of Anthropological Archaeology, 22*(1), 42–74.

Knappett, C. (2011). *An archaeology of interaction: Network perspectives on material culture and society*. Oxford: Oxford University Press.

Knappett, C. (2018). The weakness of strong ties? Communities of practice and network dynamics in the Bronze Age Aegean. *Journal of Archaeological Method and Theory, 25*(4), 974–995.

Lave, J., & Wenger, E. (1991). *Situated learning: Legitimate peripheral participation*. Cambridge: Cambridge University Press.

Levy, T. E. (1995). Cult, metallurgy and rank societies–Chalcolithic period (ca. 4500–3500 BCE). In T.E. Levy (Ed.), *The archaeology of society in the Holy Land* (pp. 226–244). London: Leicester University Press.

Levy, T. E., & Holl, A. (1988). Les sociétés chalcolithiques de la Palestine et l'émergence de chefferies. *Archives Européennes de Sociologie, XXIX*, 283–316.

Levy, T. E., Burton, M. M., & Rowan, Y. M. (2006). Chalcolithic hamlet excavations near Shiqmim, Negev desert, Israel. *Journal of Field Archaeology, 31*(1), 41–60.

Lovell, J. L., & Rowan, Y. M. (Eds.). (2011). *Culture, chronology and the Chalcolithic: Theorie and transition*. Oxford: Oxbow Books.

Manem, S. (2008). *Etude des fondements technologiques de la culture des Duffaits (âge du Bronze moyen)*. Nanterre: University Paris-Nanterre.

Manzo, G. (2007). Variables, mechanisms, and simulations: Can the three methods be synthesized? *Revue Française de Sociologie, 48*(5), 35–71.

Manzo, G. (2014). Data, generative models, and mechanisms: More on the principles of analytical sociology. In G. Manzo (Ed.), *Analytical Sociology. Actions and Networks* (pp. 4–52). Chichester, UK: John Wiley & Sons.

Manzo, G., Gabbriellini, S., Roux, V., & M'Mbogori, F. N. J. (2018). Complex contagions and the diffusion of innovations: Evidence from a small-N study. *Journal of Archaeological Method and Theory, 25*(4), 1109–1164.

Mayor, A. (2010). Ceramic traditions and ethnicity in the Niger Bend, West Africa. *Ethnoarchaeology, 2*(1), 5–48.

Mesoudi, A. (2009). How cultural evolutionary theory can inform social psychology and vice versa. *Psychological Review, 116*(4), 929.

Mesoudi, A., & O'Brien, M. J. (2009). Placing archaeology within a unified science of cultural evolution. In S. J. Shennan (Ed.), *Pattern and process in cultural evolution* (pp. 21–32). Berkeley and Los Angeles, CA: University of California Press.

Mills, B. J., Clark, J. J., Peeples, M. A., Haas, W. R., Roberts, J. M., Hill, J. B., et al. (2013). Transformation of social networks in the late pre-Hispanic US Southwest. *Proceedings of the National Academy of Sciences, 110*(15), 5785–5790.

Neiman, F. D. (1995). Stylistic variation in evolutionary perspective: Inferences from decorative diversity and interassemblage distance in Illinois Woodland ceramic assemblages. *American Antiquity, 60*(1), 7–36.

O'Brien, M. J., & Bentley, R. A. (2011). Stimulated variation and cascades: Two processes in the evolution of complex technological systems. *Journal of Archaeological Method and Theory, 18*(4), 309–337.

O'Brien, M. J., Darwent, J., & Lyman, R. L. (2001). Cladistics is useful for reconstructing archaeological phylogenies: Palaeoindian points from the southeastern United States. *Journal of Archaeological Science, 28*(10), 1115–1136.

O'Brien, M. J., Lyman, R. L., Glover, D. S., & Darwent, J. (2003). *Cladistics and archaeology*. Salt Lake City: University of Utah Press.

O'Brien, M. J., Lyman, R. L., Mesoudi, A., & VanPool, T. L. (2010). Cultural traits as units of analysis. *Philosophical Transactions of the Royal Society of London B: Biological Sciences, 365*(1559), 3797–3806.

Östborn, P., & Gerding, H. (2014). Network analysis of archaeological data: A systematic approach. *Journal of Archaeological Science, 46*(2), 75–88.

Östborn, P., & Gerding, H. (2015). The diffusion of fired bricks in Hellenistic Europe: A similarity network analysis. *Journal of Archaeological Method and Theory, 22*(1), 306–344.

Perlès, C. (2013). Tempi of change: When soloists don't play together. Arrhythmia in 'continuous' change. *Journal of Archaeological Method and Theory, 20*(2), 281–299.

Powell, A., Shennan, S. J., & Thomas, M. G. (2010). Demography and variation in the accumulation of culturally inherited skills. In M. J. O'Brien & S. J. Shennan (Eds.), *Innovation in cultural systems. Contributions from evolutionary anthropology* (pp. 137–160). Cambridge, London: The MIT Press.

Rogers, E. M. (1962). *Diffusion of Innovations*. New York: Free Press.

Rosen, S. A. (1997). *Lithics after stone age. A handbook of stone tools from the Levant*. Walnut Creek, London, New Delhi: Altamira Press.

Rosenberg, D., & Shimelmitz, R. (2017). Perforated stars: Networks of prestige item exchange and the role of perforated flint objects in the late Chalcolithic of the Southern Levant. *Current Anthropology, 58*(2), 295–306.

Rosenberg, D., Chasan, R., & van den Brink, E. C. (2016). Craft specialization, production and exchange in the Chalcolithic of the southern Levant: Insights from the study of the basalt bowl assemblage from Namir Road, Tel Aviv, Israel. *Euroasian Prehistory, 13*, 1–23.

Roux, V. (2003). A dynamic systems framework for studying technological change: Application to the emergence of the potter's wheel in the southern Levant. *Journal of Archaeological Method and Theory, 10*(1), 1–30.

Roux, V. (2010). Technological innovations and developmental trajectories: Social factors as evolutionary forces. In M. J. O'Brien & S. J. Shennan (Eds.), *Innovation in cultural systems. Contributions from evolutionary anthropology* (pp. 217–234). Cambridge, London: The MIT Press.

Roux, V. (2015). Standardization of ceramic assemblages: Transmission mechanisms and diffusion of morpho-functional traits across social boundaries. *Journal of Anthropological Archaeology, 40*(4), 1–9.

Roux, V. (2016). *Des Céramiques et des Hommes. Décoder les assemblages archéologiques*. Nanterre: Presses Universitaires de Paris Ouest.

Roux, V. (2019). The Ghassulian ceramic tradition: A single *chaîne opératoire* prevalent throughout the Southern Levant. *Journal of Eastern Mediterranean Archaeology and Heritage Studies, 7*(1), 23–43.

Roux, V., & Courty, M. A. (2005). Identifying social entities at a macro-regional level: Chalcolithic ceramics of South Levant as a case study. In A. Livingstone Smith, D. Bosquet, & R. Martineau (Eds.), *Pottery manufacturing processes: Reconstruction and interpretation* (pp. 201–214). Oxford: BAR International Series.

Roux, V., & Courty, M. A. (2007). Analyse techno-pétrographique céramique et interprétation fonctionnelle des sites: un exemple d'application dans le Levant Sud Chalcolithique. In A. Bain, J. Chabot, & M. Mousette (Eds.), *Recherches en archéométrie: la mesure du passé* (pp. 153–167). Oxford: Archaeopress.

Roux, V., Bril, B., Cauliez, J., Goujon, A. L., Lara, C., de Saulieu, G., & Zangato, E. (2017). Persisting technological boundaries: Social interactions, cognitive correlations and polarization. *Journal of Anthropological Archaeology, 48*(4), 320–335.

Rowan, Y. M., & Golden, J. (2009). The Chalcolithic period of the Southern Levant: A synthetic review. *Journal of World Prehistory, 22*(1), 1–92.

Shennan, S. J. (2001). Demography and cultural innovation: A model and its implications for the emergence of modern human culture. *Cambridge Archaeological Journal, 11*(1), 5–16.

Shennan, S. J. (2002). *Genes, memes and human history: Darwinian archaeology and cultural evolution*. London: Thames & Hudson.

Shennan, S. (2013). Lineages of cultural transmission. In E. Roy, S. J. Lycett, & S. E. Johns (Eds.), *Understanding cultural transmission in anthropology: A critical synthesis* (pp. 346–360). Oxford: Berghahn Books.

Shennan, S. J., & Steele, J. (1999). Cultural learning in hominids: A behavioural ecological approach. In H. Box & K. Gibson (Eds.), *Mammalian social learning. Symposia of the zoological society of London 70* (pp. 367–388). Cambridge: Cambridge University Press.

Shennan, S. J., & Wilkinson, J. R. (2001). Ceramic style change and neutral evolution: A case study from Neolithic Europe. *American Antiquity, 66*(4), 577–593.

Shennan, S. J., Crema, E. R., & Kerig, T. (2015). Isolation-by-distance, homophily, and "core" vs. "package" cultural evolution models in Neolithic Europe. *Evolution and Human Behavior, 36*(2), 103–109.

Stark, M. T. (Ed.). (1998). *The archaeology of social boundaries*. Washington, London: Smithsonian Institution Press.

Stark, M. T., Bishop, R. L., & Miska, E. (2000). Ceramic technology and social boundaries: Cultural practices in Kalinga clay selection and use. *Journal of Archaeological Method and Theory, 7*(4), 295–332.

Stark, M. T., Bowser, B. J., & Horne, L. (Eds.). (2008). *Cultural transmission and material culture. Breaking down boundaries*. Tucson: The University Arizona Press.

Tehrani, J. J., & Collard, M. (2009). On the relationship between interindividual cultural transmission and population-level cultural diversity: A case study of weaving in Iranian tribal populations. *Evolution and Human Behavior, 30*(4), 286–300.

Tehrani, J. J., & Riede, F. (2008). Towards an archaeology of pedagogy: Learning, teaching and the generation of material culture traditions. *World Archaeology, 40*(3), 316–331.

Valente, T. W. (1996). Social network thresholds in the diffusion of innovations. *Social Networks, 18*(1), 69–89.

Valente, T. W. (1999). *Network models of the diffusion of innovations*. Cresskill: Hampton Press.

Wenger, E. (2000). Communities of practice and social learning systems. *Organization, 7*(2), 225–246.

Ethnoarchaeology-Based Modelling to Investigate Economic Transformations and Land-Use Change in the Alpine Uplands

Francesco Carrer, Graeme Sarson, Andrew Baggaley, Anvar Shukurov, and Diego E. Angelucci

Introduction

Environmental and morphological characteristics of mountain regions provide significant constraints to pre-industrial subsistence strategies. The inverse correlation between temperature and elevation (temperature decreases as the elevation increases) affects the duration of vegetative period and the nature of the soil. Consequently, agriculture yields tend to decrease as the elevation increases, thus conditioning land-use strategies. Furthermore, terrain variability in mountain regions has an impact on mobility and production costs, influences slope instability and reduces the available arable land. All these constraining factors suggest that human groups had to develop specific adaptive strategies to survive in these extreme environments (Netting 1981). On the other hand, the aforementioned environmental and morphological characteristics contribute to amplify the consequences of human impact on the vulnerable mountain ecosystems (Previtali 2011). The carrying capacity in the mountains is much lower than in other areas, and overexploitation of these environments leads to biomass reduction, decrease in biodiversity, soil-loss and hydrogeological instability, with dramatic effects on local landscapes, ecosystem services provision and downstream areas. Therefore, understanding socioecological processes in mountain areas is of key importance not only for unravelling human

---

F. Carrer (✉)
McCord Centre for Landscape, School of History Classics and Archaeology, Newcastle University, Newcastle upon Tyne, UK
e-mail: francesco.carrer@newcastle.ac.uk

G. Sarson · A. Baggaley · A. Shukurov
School of Mathematics Statistics and Physics, Newcastle University,
Newcastle upon Tyne, UK

D. E. Angelucci
Dipartimento di Lettere e Filosofia, Università degli Studi di Trento, Trento, Italy

resilience and adaptability in extreme environments but also for identifying and promoting sustainable land-use strategies. Premodern practices and ecological knowledge of mountain communities can be investigated using documentary sources and ethnographic methods (Mathieu 2009; Netting 1981), whereas the prehistoric or early-historic origin of these practices and knowledge is inferred from archaeological data and palaeoecological proxies (Schmidl and Oeggl 2005).

Human-Environment Interaction in the Uplands

Archaeologists, ethnohistorians and palaeoecologists are increasingly interested in the long-term interaction of small-scale human communities with the alpine and subalpine zones of mountain ecosystems (corresponding to the open pasturelands above the timberline). Here the vegetative period is too short to make crop cultivation cost-effective, and therefore they are traditionally associated with mobile pastoralism and other seasonal farming activities. In the Western European mountain regions, the earliest occupation of the high altitudes is archaeologically documented since the end of the Ice Age, when human groups exploited the mountain grasslands for hunting (Fontana and Visentin 2016). Pastoral colonisation of the uplands began during the Neolithic and expanded and intensified during the Bronze and Iron Age, stimulated by population growth, increasing sociocultural complexity and new productive and economic strategies (Gleirscher 2010; Walsh et al. 2014; Carrer et al. 2016; Reitmaier et al. 2017). Besides, in the late-prehistoric period, other high-altitude activities, like ore-mining, started to coexist along with pastoralism (Bourgarit et al. 2008). Historical periods saw a further intensification of rural and non-rural upland practices, with the pinnacle corresponding to the early-modern period, and a steep decrease in the last centuries, related to a series of interweaved factors: demographic pressure, crisis and collapse of small-scale rural economies, super-regional political and economic strategies, etc. (Rosenberg 1988). This rapid overview shows quite clearly that adaptation to ecological conditions is not the only parameter to consider while analysing socioecological dynamics at high altitude. Transformations in the political background, demographic fluctuations and economic upheavals had a critical role in shaping human seasonal strategies, at least since the late-prehistoric periods.

On the other hand, the aforementioned long-term evolution of human strategies had profound consequences on the upland environments. Current mountain landscapes, albeit looking natural and pristine, are the results of millennia of direct or indirect human intervention: from animal grazing, to forest clearing, to the construction of shelters, animal enclosures, terraces, canals for irrigation and paths (Catalan et al. 2017). These pre-industrial practices are generally perceived as perfectly adapted to the characteristics of mountain environments and are assumed to be ecologically sustainable. However, recent research has shown that this has not always been the case and that the history of human interaction with high-altitude environments is often characterised by overestimations of the carrying capacity of an ecosystem, misperception of its vulnerability and underestimation of the long-term consequences (Brisset et al. 2017). Sustainability cannot be taken for granted for premodern upland strategies, but it needs to be specifically evaluated.

Assessing adaptability and sustainability of premodern high-altitude practices requires an in-depth understanding of the traditional ecological knowledge, which can be reasonably acquired only for the last few centuries (Fernández-Giménez and Fillat Estaque 2012). The resolution of archaeological and palaeoecological data prevents a detailed reconstruction of ancient strategies, and the evaluation of human-environment interaction at high altitude in prehistoric and early-historic periods is usually limited to a generic assessment of their intensity. Therefore, the estimation of the effect of specific economic practices on upland ecosystems neglects the early phases of development of these practices, affecting the estimation of their long-term sustainability and resilience. This issue is not only limited to mountain areas and high-mountain activities; indeed ecologists all around the world are increasingly aware of the scientific limitations associated to it (Bennett et al. 2015). Two approaches have been identified by the scientific community as the most suitable for addressing the problem: the study of the relationship between humans and material culture in the present to create analogical models or narratives for the past (ethnoarchaeology) and the mathematical simulation of alternative scenarios given specific ecological and behavioural constraints (computer modelling) (Balbo et al. 2016).

Ethnoarchaeological Approaches to Land-Use and Landscape Transformation

Ethnoarchaeology is a subdiscipline of archaeology that studies the material results of human behaviour: from the production, use and discard of artefacts, to the relationships with animals and plants, to the management of intra-site and inter-site spaces (David and Kramer 2001). The original purpose of ethnoarchaeology was to provide analogical inferences to interpret archaeological records, assemblages and landscapes. In the last decades, this subsidiary role has been questioned, and ethnoarchaeology has increasingly been perceived as a standalone discipline that can provide useful insight to understand current world and especially pre-industrial and marginalised communities (Cunningham 2009; Skibo 2009). However, the use of ethnoarchaeology for analogical purposes has not been completely abandoned. Recent developments in quantitative archaeology have stimulated the use of ethnoarchaeological case studies to test and calibrate the statistical and mathematical tools used in archaeological research. This experimental approach, well established in the study of material culture, has recently developed also for the spatial analysis of inhabited contexts (Lancelotti et al. 2017). Ethnographic and ethnoarchaeological inferences are also increasingly used to understand human-environment interaction in the past. Adaptation strategies of small-scale societies and the ecological impact of pre-industrial practices are investigated in the modern world, informing not only future development and environmental management policies but also the interpretation of prehistoric/early-historic socioecological systems and landscapes (Biagetti 2014).

Mountain regions represent an important reservoir of pre-industrial rural strategies in Europe, since their aforementioned characteristics have prevented a full industrialisation of farming and land-use intensification. Since the twentieth century, ethnographic and ethnohistorical research have investigated different aspects of the economy and society of mountain communities, in particular in the Alpine region (Cole and Wolf 1974; Netting 1981; Rosenberg 1988; Viazzo 1989). Recent projects have particularly focused on the transformation of traditional land use, farmer knowledge and landscape identities in mountain areas (Rescia et al. 2008; Dossche et al. 2016). Upland rural activities, like transhumance, summer dairy and haymaking, have also been widely analysed (for the Alps, see, e.g. Garde et al. 2014; Viazzo and Woolf 2001). Despite the remarkable importance of ethnography and ethnohistory in the European mountains, ethnoarchaeological research is instead quite rare. Some projects in the 1980s and early 1990s were aimed at providing interpretative tools to archaeologists investigating prehistoric mountain strategies (Barker and Grant 1991; Chang and Tourtellotte 1993). Although the use of ethnoarchaeological analogy has not disappeared (Carrer 2015; Le Couédic 2012), current projects are more focused on the identification of recent patterns of landscape and land-use change for informing current and future policy (Christie et al. 2007; Mientjes 2004). The rapid disappearance of pre-industrial rural practices, and the growing interest of archaeologists and palaeoecologists for human interaction with mountain environments (especially the high mountains), requires an expansion and intensification of mountain ethnoarchaeology in the next decades (Carrer et al. 2015).

Ethnoarchaeological Inferences for Modelling Landscape and Land-Use Change

The use of computational models for simulating past socioecological systems is known and well established in archaeology (Barton et al. 2010; Lake 2014) and is even more popular in other fields, like quantitative ecology (Filatova et al. 2013; Dearing et al. 2010). Alternative scenarios are created by integrating different ecological processes and human decision-making rules, in order to test specific hypothesis related to the sustainability and resilience of (pre)historic human-environment interactions (Danielisová et al. 2015). The creation of decision-making rules for simulation usually relies on reconstructions of past behaviours based on archaeological evidence. But quantitative archaeologists are increasingly aware of the limitations associated to this approach, such as the circularity of the argument (inferring behavioural rules from archaeological interpretation to improve the archaeological interpretation) and the equifinality (different behaviours can lead to similar archaeological evidence). Ethnographic observations are increasingly employed in computer modelling to infer past behaviour and are particularly common in the study of hunter-gatherer society and ecology (Barceló et al. 2015; Briz i Godino et al. 2014). This approach consists in extrapolating socio-economic strategies observed in small-scale communities and use them to infer the simulation parameters. Consequently, the use of ethnography in archaeological simulation faces the same theoretical challenges associated with ethnographic analogy in archaeological interpretation, widely discussed in theoretical works during the 1980s (Orme 1981; Wylie 1982): the unsystematic and question-driven use of behavioural strategies, isolated from their economic, cultural and social context, might lead to inaccurate estimations of past behaviours.

In this paper we promote an alternative and more solid approach to ethnographic analogy: the application of computer simulation to ethnoarchaeological case studies. Decision-making rules are historically and ethnographically analysed in a pre-industrial context and used to create a simulation scenario to be applied in the same context. If the results of the simulation differ considerably from the observed status of the investigated context, the simulation parameters need to be calibrated or additional parameters need to be introduced. As pointed out before, this 'experimental' approach to ethnoarchaeological analogy has proved very effective in modelling intra-site and inter-site spatial patterns (Surovell and O'Brien 2016; Rondelli et al. 2014; Biagetti et al. 2016; Carrer 2013; Carrer 2017), but to the best of our knowledge, it is still unknown in archaeological simulation.[1]

In order to explore the potentials of ethnoarchaeology-based computer simulation, we have used a static mathematical model to analyse rural landscape and land-use change in two small villages of the Italian Alps during the eighteenth–nineteenth centuries. Particular attention is paid to the transformations that occurred in the high-altitude sector of the analysed territory. In this area, like in many other areas of the region, the seasonally exploited upland landscapes underwent a rapid and abrupt reconfiguration, which is commonly associated with a significant economic shift which occurred in the nineteenth century: from unspecialised seasonal pastoralism to dairy-focused cattle transhumance (Carrer and Angelucci 2017). Historical reconstructions, in turn, attribute this economic change (and the consequent landscape change) to demographic pressure and macroeconomic processes (like the decreasing value of wool) (Mathieu 2009). Therefore, the main questions addressed by this ethnoarchaeological model are as follows: (1) Is demographic fluctuations a key driver of economic change in mountain rural communities? (2) Do macroeconomic dynamics influence local subsistence transformations? (3) Does local economic change determine land-use change at high altitude? (4) Does a more specialised farming economy affect ecological sustainability? The outcomes of the model will provide interesting data to understand socioecological processes in Val di Sole during the transition to

---

[1] Luke Premo introduced the concept of 'experimental ethnoarchaeology' to define the use of alternative theoretical scenarios in agent-based modelling to test specific research hypotheses (Premo 2007). However, this approach does not include the application of computer modelling to living communities and is conceptually closer to experimental archaeology.

modernity, but they will also provide critical insights to investigate the earlier evolution of rural communities in mountain regions, their economic resilience and their ecological sustainability.

Historical Evolution of the Upland Landscapes of Val di Sole (Italian Alps)

The area selected for this study corresponds to the rural landscapes pertaining to two small villages located on the northern (south-facing) slope of Val di Sole, in the Trentino region (Italian Alps): Ortisé and Menas (Fig. 1). The upland sector of their territory has been subjected to intensive archaeological, geoarchaeological and ethnoarchaeological research since 2010, complemented by the analysis of local historical sources. These investigations shed new light on the development of high-altitude human-environment interactions in this area and contributed to improving our understanding of long-term socioecological dynamics in the Alps.

Study Area: Val di Sole and the Villages of Ortisé and Menas

Val di Sole is a ca. 40-km-long Alpine valley located in the Italian region of Trentino. Its WSW-ENE orientation is responsible for a marked asymmetry between its side slopes as far as microclimate, geomorphology and land use are concerned. Permanent villages are situated along the valley bottom, and none is found on the south (north-facing) slope; only a few *malghe* (seasonal dwellings exploited by herders for livestock grazing and cheese production) do exist at higher elevation in the uplands of the south slope. In contrast, permanently inhabited villages are found on natural terraces up to ca. 1500 m elevation on the north (south-facing) slope – this is the case of both Ortisé and Menas (Fig. 2f–g). Several *malghe* are scattered on the uplands of the North Slope, and mountain pastures reach ca. 2500 m height. This setting causes traditional land use in Val di Sole to be arranged on a 'vertical' basis, with seasonal movements to and from the valley bottom and the uplands, especially for activities related to animal husbandry.

The community of Ortisé and Menas owns the property and the right of use of the stretch of land upslope from the villages. The territory includes productive forests (mostly conifers) up to 2000–2200 m elevation, extensive pastures/grassland between ca. 2000 and 2500 m and unproductive land at higher elevation (mostly rock outcrops and bare glacial or periglacial morphologies). Winter stables for animals are located at short distance from the villages, at around 1600 m elevation, while *malghe* are found all the way up and reach the elevation of 2200 m.

Geologically, the upland area of Ortisé and Menas is made up of metamorphic rocks, mostly gneiss, locally covered by glacial and periglacial deposits and often

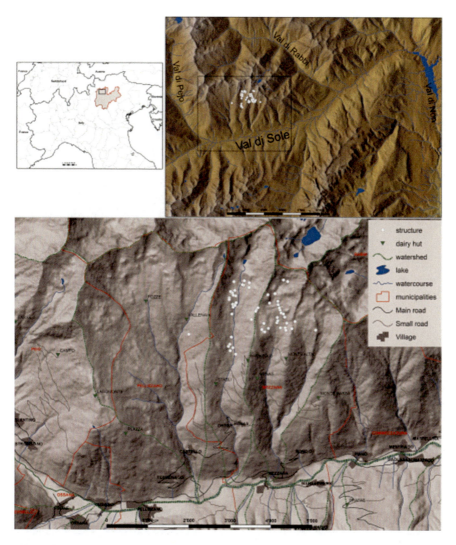

Fig. 1 Map of the study area

affected by slope and mass movements. The climate of the upland is Alpine, cold and humid, with average annual mean temperature around 0 °C, annual precipitation of ca. 1000 mm/a and about a half of the annual cycle dominated by frost and snow. Geological and climate constraints control both hydrography and soil: water is abundant all-year round, and soils are mostly thin and desaturated, even if human impact has changed their properties in the most intensively grazed pastures, due to the indirect input of organic matter and animal dung over centuries of pastoral exploitation (see Angelucci et al. 2014 for details).

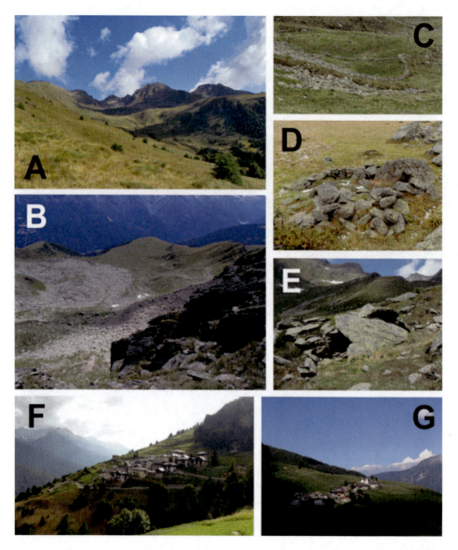

Fig. 2 Landscapes of the study area: Val Molinac (**a**) from the south and Val Poré (**b**) from the north; example of pastoral enclosure (**c**), dry-stone hut (**d**) and rock shelter (**e**); the villages of Ortisé (**f**) and Menas (**g**)

Landscape Archaeology in Val Molinac and Val Poré

The two upland valleys within the territory of Ortisé and Menas, known as Val Molinac and Val Poré, have been the main study area of the "ALPES" ("Alpine Landscapes: Pastoralism and Environment of Val di Sole") research project since 2010. The aim of the project is to study the traces left by pastoral exploitation in the uplands of Val di Sole under a diachronic and archaeological perspective and to understand the mutual relationships between natural and cultural factors in the evolution of the mountain lands (see Angelucci and Carrer 2015).

Archaeological surveys and excavations have identified several relict landscape features related to the historical exploitation of the land, scattered in both valleys between ca. 1900 and 2450 m elevation. They are mostly dry-stone structures related to pastoral use (Fig. 2a–e). Numerous huts and rock shelters were recorded, but the most noticeable structures are enclosures, used as pens to gather the livestock. Several small- to medium-size enclosures (up to few hundred square metres) were identified in the study area, as well as a few large ones (up to 1000 square metres), both single and compound. Their characteristics are rather varied, and assessing the age of their construction and use is often difficult. Archaeological excavation and test pits undertaken in large and compound enclosures showing similar characteristics (such as shape, position in the landscape, building technique and average size) have yielded quite homogeneous archaeological evidence. Most of the large and compound enclosures turned out to be built in late-medieval and early-modern times (fifteenth to seventeenth centuries AD – see Dell'Amore et al. 2017) and used with variable intensity until recent times (mid-twentieth century AD, as reported by local informants). Still, archaeological data indicate earlier phases of human occupation of the area, namely, during the Bronze Age (second millennium BC), between the Late Bronze Age and the Early Iron Age (twelfth–eighth centuries BC) and in earlier stages of the Middle Ages (see Angelucci and Carrer 2015 and Angelucci et al. 2017). Archaeological research has shown that the uplands of Ortisé and Menas have been seasonally exploited for distinct purposes for long time and that the late-medieval to early-modern features mark the formation of a landscape, progressively oriented towards the intensive, structured exploitation of the uplands for pastoral production.

Historical Background (Fifteenth–Twentieth Centuries)

Rural Alpine communities, like Ortisé and Menas, underwent profound socio-economical transformations between the fifteenth and twentieth centuries. Economic historians explain this upheaval as a consequence of two interplaying factors: demographic increase and macroeconomic trends at continental level (Mathieu 2009). The Alpine region experienced a steady but significant population growth in the early-modern period. Higher population density and the limited arable surface in mountain areas triggered a series of correlated processes: more intensive farming activities led to a higher human impact on the environment, which in turn caused loss of crucial ecosystem services (soil, biodiversity, etc.). Local strategies to lower the pressure on natural resources included non-permanent emigration and higher reliance on animal products, to exploit more effectively the marginal areas (including the uplands). As a consequence of this, overgrazing became a major issue in many parts of the Alps. External factors contributed to influencing the aforementioned changes as well. The wool market grew in Europe between the sixteenth and the eighteenth century and persuaded local authorities to promote sheep husbandry (see Varanini 2004 for the Trentino region). On the other hand, new crops were introduced in the Alpine region: buckwheat from the fifteenth century, corn during the sixteenth century and potato from the nineteenth century. It is worth noticing, though, that in the Ortisé and Menas area, the predominant crops remained wheat and barley until the introduction of potato (Castiglioni 1976).

During the eighteenth century, the price of wool decreased dramatically, under the pressure of the cheaper wool from East Asia. Therefore, local authorities started shifting their economic interest towards dairy produce. Since the early nineteenth century, incentives were given to farmers all around the Alps to replace small livestock (sheep and goats) with cattle and to focus their economic strategies on the production of butter and cheese for the market. This initiative was mainly aimed at improving the economic conditions of Alpine communities and promoting a more effective exploitation of natural resources in the mountains (in this period forest and pasture degradation was simplistically attributed to the inadequacy of traditional practices and ecological knowledge: see Battisti 1904 for the Trentino region). Meanwhile, increasing population pressure prompted the transition from seasonal to permanent emigration, thus rapidly reversing the demographic process. At the beginning of the twentieth century, the Alpine region was economically marginal and increasingly depopulated, and rural communities were largely relying on extensive pastoralism and new industrial activities (e.g. mining or tannin extraction) (Rosenberg 1988).

Pre-industrial Pastoral and Farming Practices in Val di Sole: An Ethnoarchaeological and Ethnohistorical Investigation

Unstructured interviews of retired farmers and herders of Ortisé enabled the collection of detailed information about farming and pastoral practices predating the crisis of local rural economy in the 1960s. The main goal of the interviews was understanding historical land-use change and the use and abandonment of the dry-stone structures documented in the high pastures. Participant observation was carried out with sheep and cattle herders, to better comprehend current practices of environmental management at high altitude. Ethnohistorical data about local economy and land use in the nineteenth and twentieth century were acquired from published research (e.g. Castiglioni 1976). The integration of ethnohistorical data complemented the information provided by archaeological research and documentary surveys. This contributed to investigating the evolution of pastoral strategies in the last three centuries in Val Molinac and Val Poré and its consequences for local upland landscapes.

During the early-modern period, until the end of the eighteenth century, local pastoralism was largely nonspecialised. Meat and dairy products were used to complement the local diet, mainly relying on cereals (rye and barley) and marginally on horticulture. The growing importance of the wool market boosted the number of sheep, at the expenses of goats and cattle. Livestock was stabled in the villages during the winter, and the low- and mid-altitude meadows, as well as the woodlands, were largely used for grazing during springtime and autumn. During the summer, sheep, goats and cattle were taken to the upland pastures for 3 months, and meadows were mowed for winter fodder. Three large dry-stone enclosures were exploited in each upland valley (Val Molinac and Val Poré) during the summer, to gather the animals for the night and probably to milk them. Dairy production took place primarily in the uplands, inside small dry-stone huts adjacent to the aforementioned enclosures. High-altitude pastures exceeded the needs of the local communities, and they were often rented to professional transhumant shepherds from the neighbouring Lombardy region (in that period under the control of the Republic of Venice).

The decreasing value of wool on the market, together with the growing demand for dairy products, triggered the transition of the local economy towards a dairy-focused cattle husbandry. Besides, a steep increase in population between the eighteenth and the nineteenth century is assumed to have exceeded the carrying capacity of local mountain environments. These factors had a profound impact on the upland landscape structure: the enclosures were replaced by more specialised compounds called *malghe*, constituted of a dairy structure adjacent to a barn. Some of these *malghe* are still in use nowadays, for pastoral purposes, whereas the enclosures have been marginally exploited until the mid-twentieth century. The larger number of cattle led to significant changes in land use, even around the villages. The longer winter-stabling period of cattle, for instance, required more fodder, and consequently a significant portion of arable land was converted to meadow. In addition, larger sectors of the high altitudes were exploited for haymaking, as suggested by the number of modern dry-stone huts recorded in these areas. The introduction of potato from the early nineteenth century represented a revolutionary change in local subsistence (Zaninelli 1979). Fallow was not practised in these mountain agro-systems, and three types of crops (rye, barley and potato) were annually rotated (Castiglioni 1976). This economic strategy characterised Ortisé and Menas, with minor variations, until the mid-twentieth century, when the increasing depopulation and the changed macroeconomic and social dynamics in Europe, Italy and the region (Trentino) led to a progressive collapse of rural economy. Today crop cultivation has almost disappeared, and all the fields are dedicated to fodder production. Two rural activities are still surviving: seasonal pastoralism and horticulture. However, they are currently considered economically and socially marginal for the local communities.

Reconstructing Upland Economic Transformations and Land-Use Change Using Ethnoarchaeologically Informed Mathematical Simulation

Ethnographic, historic and archaeological information were used to create formalised behavioural rules, in turn employed to produce realistic socio-economic scenarios and simulate pre-industrial subsistence of small-scale mountain

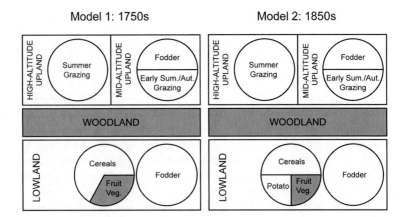

Fig. 3 Schematic representation of land-use simulation scenarios in the 1750s and 1850s

communities. Two static mathematical models were developed to analyse two different snapshots in the historical evolution of Alpine rural economy (Fig. 3). The first model includes demographic parameters, farming practices and economic strategies that approximate the lifeways in Ortisé and Menas around the 1750s. The second model delineates the socioecological system of Ortisé and Menas in the 1850s, after the introduction of new crops and the transition towards a more intensive pastoral economy. The mathematical structure of the models was adapted from an existing analytical protocol, recently developed by some of the authors to reconstruct late-prehistoric subsistence strategies in Eastern Europe (Shukurov et al. 2015). The results of computer simulations in the two study areas were tested against the historic land-use patterns described by ethnographic and documentary sources, in order to assess the reliability of the models and inspect the potential inconsistencies.

Parameters

The 1750s model (Model 1 = M1) describes a local subsistence system relying primarily on cereal cultivation and animal husbandry. The latter is characterised by short-scale seasonal mobility and a weak specialisation on wool trading. Historical sources suggest that cereals had a higher importance on the local diet than domestic animal products (including dairy products), whereas wild animals contributed only marginally to the calorific intake of local inhabitants. The 1850s model (Model 2 = M2), on the other hand, delineates a scenario where subsistence agriculture is progressively replaced by intensive and dairy-focused cattle husbandry. This transition produces a radical economic transformation, with an increasing economic importance of cheese trading. Table 1 describes in detail the different parameter selected for describing the economy and ecology of Ortisé and Menas between the

Table 1 Model parameters

Category	Parameter	Name	M1	Scopenote M1	M2	Scopenote M2
Calorific fraction of diet	Cereal products	ϵ_g	0.55	Inferred from Castiglioni (1976), Battisti (1904)	0.2316	Inferred from Castiglioni (1976), Battisti (1904)
	Domestic animal products	ϵ_d	0.35	Battisti (1904)	0.37	Battisti (1904)
	Wild animal products	ϵ_w	0.10	Estimated	0.06	Estimated
	Potatoes	ϵ_p	0	Not cultivated (Castiglioni 1976)	0.3384	Inferred from Castiglioni (1976)
Domestic animals parameters	Fraction of cattle	a_c	0.25	Inferred from Battisti (1904)	0.40	Battisti (1904)
	Fraction of sheep/goats	a_s	0.625		0.55	
	Fraction of pigs	a_{pi}	0.125		0.05	
	Usable meat: cattle (kg/hd)	m_c	100	Inferred from Shukurov et al. (2015)	100	Inferred from Shukurov et al. (2015)
	Usable meat: sheep/goat (kg/hd)	m_s	25		25	
	Usable meat: pig (kg/hd)	m_{pi}	20		20	
	Energy content: cattle (kcal/kg)	e_c	1600		1600	
	Energy content: sheep/goat (kcal/kg)	e_s	1600		1600	
	Energy content: pig (kcal/kg)	e_p	3000		3000	
	Fraction of cattle culled annually	k_c	0.2	Inferred from Cribb (1984)	0.4	Inferred from Cribb (1984)
	Fraction of sheep/goat culled annually	k_s	0.2		0.2	
	Fraction of pigs culled annually	k_{pi}	0.5		0.5	
	Fraction of milking cows in a cattle herd	k_c	0.4	Inferred from Battisti (1904)	0.55	Battisti (1904)
	Fraction of milking ewes/does in a flock	k_s	0.2		0.3	
	Surplus cow milk after weaning (l/yr/hd)	y_c	1150	~2000 l/yr/hd milk prod. (0.4 modern alpine prod.); calves intake ~5–6 l/day/hd; weaning period ~7–8 months (World Bank)	2500	~3000 l/yr/hd milk prod. (0.6 modern alpine prod.); calves intake ~5–6 l/day/hd; weaning period ~2–3 months (World Bank)
	Surplus sheep/goat milk after weaning (l/yr/hd)	y_s	225	Average ~100 l/yr/hd sheep + alpine goats ~500 l/yr/hd; lambs/kids intake ~0.6 l/day/hd; weaning ~5 months (World Bank)	250	Average ~100 l/yr/hd sheep + alpine goats ~500 l/yr/hd; lambs/kids intake ~0.6 l/day/hd; weaning ~2 months (World Bank)

(continued)

Table 1 (continued)

Category	Parameter	Name	M1	Scopenote M1	M2	Scopenote M2
Grazing and foddering areas	Grazing area: cattle (ha/hd)	A_c	1	(Gregg 1988: adapted for 6 months grazing)	1	(Gregg 1988: adapted for 6 months grazing)
	Grazing area: sheep/goats (ha/hd)	A_s	0.1		0.1	
	Grazing area: pigs (ha/hd)	A_p	0	Grazing in the woodlands	0	Grazing in the woodlands
	Area for winter fodder: cattle (ha/hd)	M_c	1.5	(Gregg 1988: adapted for 6 months growing, to produce 6 months fodder)	1.5	(Gregg 1988: adapted for 6 months growing, to produce 6 months fodder)
	Area for winter fodder: sheep/goats (ha/hd)	M_s	0.05		0.05	
	Fraction of fodder as hay	$\aleph$	0.8	Estimated	0.8	Estimated
Wild animal products	Usable meat: red deer (kg/hd)	m_r	130	Inferred from Shukurov et al. (2015)	130	Inferred from Shukurov et al. (2015)
	Usable meat: roe deer (kg/hd)	m_{ro}	11		11	
	Usable meat: wild boar (kg/hd)	m_b	130		130	
	Usable meat: chamois (kg/hd)	m_{ch}	20		20	
	Usable meat: ibex (kg/hd)	m_i	35		35	
	Fraction of red deer among wild animals	a_r	0.25	Inferred from Swiss National Park and National Park of the Dolomites (Italy)	0.25	Inferred from Swiss National Park and National Park of the Dolomites (Italy)
	Fraction of roe deer among wild animals	a_{ro}	0.25		0.25	
	Fraction of wild board among wild animals	a_b	0.125		0.125	
	Fraction of chamois among wild animals	a_{ch}	0.25		0.25	
	Fraction of ibex among wild animals	a_i	0.125		0.125	
	Energy content: red deer (kcal/kg)	e_r	1400	Inferred from Shukurov et al. (2015)	1400	Inferred from Shukurov et al. (2015)
	Energy content: roe deer (kcal/kg)	e_{ro}	1400		1400	
	Energy content: wild boar (kcal/kg)	e_b	3500		3500	
	Energy content: chamois (kcal/kg)	e_{ch}	1600		1600	
	Energy content: ibex (kcal/kg)	e_i	1600		1600	

Labour productivity and calorie intake	People per family	N_f	8	Inferred from Viazzo (1989)	8	Inferred from Viazzo (1989)
	Fraction of family who work in the field	W	0.625		0.625	
	Area ploughed with ard (m²/pn-hr)	s_t	260	Inferred from Gregg (1988), Boserup (1965)	260	Inferred from Gregg (1988), Boserup (1965)
	Cereal area reaped/threshed (m²/pn-hr)	s_r	30		3	
	Grass area cut (m²/pn-hr)	s_c	60		60	
	Potato area harvested (m²/pn-hr)	s_p	–		16	
	Length of working day (hr/day)	τ	10	Estimated	10	Estimated
	Length of working year (day/yr)	σ_{yr}	250		250	
	Fraction of family requiring reduced calories (young/old)	δ_{yo}	0.5	Inferred from Viazzo (1989)	0.5	Inferred from Viazzo (1989)
	Daily energy requirement per person (kcal/pn/day)	C	2500	Estimated	2500	Estimated
	Daily energy requirement p.p. young/old (kcal/pn/day)	C_{yo}	c/2		c/2	
Agriculture parameters	Seeding fraction of the yield	γ	0.12	Estimated	0.12	Estimated
	Yield fraction lost to pest, etc.	λ	0.25		0.25	
	Energy content of crops (kcal/kg)	e_g	3820	Average of rye (3750) and barley (3890) (FAO)	3820	Average of rye (3750) and barley (3890); source: FAO
	Energy content of potato (kcal/kg)	e_p	670	(FAO)	670	(FAO)
	Fruit/vegetables fraction of land	δ_f	0.5		0.5	
	Settlement area (ha)	A_0	2	Ortisé+Menas, from cadastre	4	Ortisé+Menas, from cadastre
	Population density (pn/ha)	ρ	75.4	Set to get a tot. population of 150	75.4	Set to get a tot. population of 300
	Unmanured cereal yield (kg/ha/yr)	Y_u	1200	~1/2 the cereal yield of Austria in 1960s (The World Bank)	1200	~1/2 the cereal yield of Austria in 1960s (The World Bank)

(continued)

Table 1 (continued)

Category	Parameter	Name	M1	Scopenote M1	M2	Scopenote M2
	Unmanured potato yield (kg/ha/yr)	Y_{pu}	1.e4	Not cultivated (Castiglioni 1976)	1.e4	Estimated
	Ratio of manured to unmanured yield (kg/ha/yr)	r_{mu}	1.2	Estimated	1.2	Estimated
	Manure produced by cattle (kg/hd/yr)	m	2.5e3		2.5e3	
	Manure applied to fields (kg/ha/yr)	μ	15.e3		15.e3	
Wool and cheese parameters	Fraction of wool used for local consumption	f_{wc}	0.2	Estimated	0.8	Estimated
	Wool production (kg/hd/yr)	y_w				
	Cost of wool (Kr/lb)	R_w	20	Schmelzer (1972), Dietrich (1980)	24	Schmelzer (1972), Dietrich (1980)
	Cost of labour (Kr/day)	R_l	15		32	
	Cost of cheese (Kr/lb)	R_{ch}	3		9	
	Fraction of dairy used for local consumption	f_{dc}	0.9	Estimated	0.5	Estimated
	Conversion from Habsburg pound to kilogramme	σ_{lb}	0.560	Schmelzer (1972), Dietrich (1980)	0.560	Schmelzer (1972), Dietrich (1980)
	Kg of hard cheese produced for 1 l of milk	δ_{ch}	0.1		0.1	
Trashumance parameters	Fraction of year that mid-altitude area is used	δ_{mid}	6./12	Castiglioni (1976), Battisti (1904)	6./12	Castiglioni (1976), Battisti (1904)
	Fraction of year that high-altitude area is used	δ_{high}	3./12		3./12	

Keys: *ha* hectare, *yr* year, *hd* head (animal), *pn* person, *hr* hour, *Kr* crowns (currency), *lb* Habsburg pound (weight)

1750s (M1) and 1850s (M2). Source, calculation and accuracy of each parameter are also specified in the table. It is worth pointing out that the two models are preliminary and are not the product of a comprehensive screening of all the sources available in the area. More accurate models will be created in the future, but the key goal of this paper is to assess the feasibility and applicability of the transdisciplinary methodology developed by the authors.

Although buckwheat and maize are documented in the region at the end of the eighteenth century (I. Franceschini, com. pers.), there is no historical evidence that these cereals were cultivated in Ortisé and Menas. On the other hand, since the potato spreads in the Trentino region during the first half of the nineteenth century and is documented in the study area during the early-twentieth century, it is assumed to be already cultivated in Ortisé and Menas during the mid-nineteenth century. The importance of wild animals, already marginal in the first model, is assumed to drop in the later model, as a consequence of overhunting – widely documented in the whole Alpine region. The proportion of wild animals in the study areas is also assumed to be stable, despite the variation in the absolute number of animals.

Variation in the domestic animal parameters between the two models depends on the economic focus of husbandry in the selected periods. The grazing area and fodder amount per animal are assumed equal in the two periods, and it is attributed a general value available for comparable premodern economies (Gregg 1988; Boserup 1965). The higher economic importance of wool in the 1750s and the higher importance of cattle dairy produce in the 1850s justify the difference in bovine/ovicaprine proportion, from ~1/3 to ~2/3. Due to the progressive specialisation of husbandry, pig decreases its importance in the diet. Although detailed information about the sex/age composition of herds and flocks is not available, zooarchaeological studies (Halstead 1996; Cribb 1984) suggest that the higher value of wool (Model 1) might lead to low culling proportion and later weaning for sheep, whereas a higher importance of milk (Model 2) might determine earlier weaning and higher culling proportion for cattle. Milk surplus and the ratio of milking animals in herds/flocks were in turn influenced by the aforementioned factors.

As detailed before, land-use organisation in Ortisé and Menas shows vertical patterns (Fig. 3). The villages are located around 1500 m asl and are surrounded by a variable proportion of meadows and cultivated land (often terraced), with the ecological limit of arable land around 1650–1700 m asl. No fallow is documented in the study area, and in the twentieth century, a 3-year crop rotation (rye, barley and potato) guaranteed moderate soil regeneration (Castiglioni 1976). A remaining fraction of land is dedicated to fruit and vegetable production. The uplands, above the timberline, are exploited seasonally and located above 1900 m asl. Large and small livestock is grazed in the uplands during the summer, roughly between the 15th of June and the 15th of September (3 months), and 50% of the nutrition for livestock is provided by winter fodder harvested in the high meadows (Castiglioni 1976). The remaining 25% of the annual nutrition is assumed to come from early-summer and early-autumn grazing in the high meadows. Although marginal grazing is historically documented also in the woodlands and in the meadows and fields nearby the villages, for parsimony and dearth of detailed information, this strategy has not been considered in the models.

The hypothetical average household is composed of eight individuals: two adults, two young adults, two children and two old adults. Adults and young adults contribute equally to the labour and require an equal amount of calories. Children and old adults are assumed to do 50% of the adults/young-adults work and require 50% of the calories. These estimations are kept constant in the two models. The population of Ortisé and Menas is one of the most important parameters, and it has been given particular attention. Lacking a specific census for the two periods considered, demography has been inferred from a combination of later census and the surface of the villages. In 1910, Ortisé had 149 inhabitants, Menas 67, but the number of residents might have been slightly higher (Castiglioni 1976). Considering the negative demographic trend documented in the twentieth century, we can assume a higher number of inhabitants for the mid-nineteenth century. In the 1850s, Ortisé and Menas had a cumulative size of ~4 ha, and this suggests ~2 ha settlement size for the previous century. A constant population density of ~75 people/hectare for the two periods led to a cautionary estimation of ~150 people for the 1750s and ~300 people for the 1850s.

Simulations

The original model focussed on a central village, surrounded by concentric 'zones' (dominated by arable land, pastures and meadows, respectively); this was appropriate for a largely homogeneous (steppe) landscape and was of interest given that the time constraints of daily travel to the fields were an issue in that context. With Alpine seasonal farming, there is a clearer distinction between different regions of land – at different altitudes, and in use for different purposes at different times of the year – and daily travel is largely replaced by longer-term stays in the different regions. We therefore instead consider such regions as separate zones in the current model: the 'lowlands' (at altitudes 1500–1700 m – surrounding the village), used for arable and horticultural land, plus some meadow (haymaking) land; the mid-altitude fields (at altitudes ca. 1800–2100 m), used for spring/autumn pasture and summer meadows; and the high-altitude fields (at altitudes 2100–2500 m) used for summer pasture.

To model the transhumance, we introduce parameters δ_{mid} and δ_{high} for the fractions of the year for which the mid-altitude and high-altitude regions, respectively, can be used. Although these are left as adjustable parameters in some of the equations below, the current estimates assume $\delta_{mid} = 0.5$ and $\delta_{high} = 0.25$. In other words, the upper pasture is in use for 3 months, whereas the mid-altitude fields are used for 6 months (3 months as pasture, before and after the high-altitude fields are available; and 3 months as meadows, when the animals are in the upper fields); fodder must sustain the animals for the remaining 6 months. As noted below, further developments of the model must explore variations from these simple fractions.

Some woodland is also assumed to be present in our model: as a refuge for the wild animals which make a minor contribution to the diet, as a grazing area for pigs, as a source of fuel and as a source of leafy fodder used to supplement animals' winter diet. At present, the woodland is treated as an infinite reservoir of resources, without any attempt to calculate the precise area required – for instance, we do not currently model the use of wood for fuel – although some consistency checks are ultimately performed for the results. Geographically, the woodland would be at altitudes between the villages and the upper-altitude fields (which extend above the timberline).

Movement between the various regions is not an issue of daily concern, so the areas are essentially considered as separate 'boxes' in the model. As a consequence, the current model no longer attempts to quantify the fraction of 'unproductive' land existing alongside each region; here we just focus on the productive land.

As noted earlier, there is no evidence that a fallow system was in use in the economies discussed; instead, the farmers used a crop-rotation system. While fruit and vegetables are important for nutrients (including vitamins), they do not contribute significantly to the calorie budget. They are therefore dealt with implicitly within our current model, simply using the appropriate fraction δ_f of the total crop land (i.e. to model this crop rotation). A later model may account for the nutrient intake in more detail. These land areas are sketched in Fig. 3.

Agricultural Land Use

The land-use calculations are based on those of Shukurov et al. (2015), under the modified base parameters summarised in Table 1. Note that the grazing areas included in the table are based on that land use being relevant for 6 months of the year. Similarly, the haymaking (meadow) areas are calculated on the basis of 6 months of productivity, and the requirement that the hay produced sustains the animals for 6 months centred on the winter. This is discussed in more detail below, in the context of seasonal farming.

As well as the differences in land zones discussed above, there are some additional changes to the earlier model, although the basic scheme of calculations (working in terms of human energy consumption from various sources) is unchanged. For completeness, we reintroduce all the relevant equations here, even where no changes are required.

The possible use of surplus dairy produce for sale (discussed in more detail below) requires the introduction of the parameter f_{dc} (the fraction of dairy produce consumed locally). In terms of this, the number of livestock per person, n_a (measured in head/person: hd/pn), is given by

$$n_a = \frac{\epsilon_d C}{k_c a_c m_c e_c + k_s a_s m_s e_s + k_{pi} a_{pi} m_{pi} e_{pi} + f_{dc}\left(\kappa_c a_c y_c e_{mc} + \kappa_s a_s y_s e_{ms}\right)},$$

where we, for simplicity, neglect horses. In terms of this, the energy consumption (e.g. in calories) per person per year, E_a (measured in kcal/pn/yr) is then given by

$$E_a = n_a \left(k_c a_c m_c e_c + k_s a_s m_s e_s + k_{pi} a_{pi} m_{pi} e_{pi}\right).$$

The grazing area for the transhumance species (cattle and sheep/goats), $A_{a,\,trans}$ (measured in ha/pn), is then

$$A_{a,trans} = n_a \left(a_c A_c + a_s A_s \right).$$

In general, some separate, lowland grazing for the non-transhumance species (pigs and horses) would be required:

$$A_{a,low} = n_a \left(a_{pi} A_{pi} + a_h A_h \right),$$

but in the present model, we assume that pigs do not require dedicated field grazing ($A_{pi} = 0$) as they forage in forests and consume scraps, and as noted above, we neglect horses ($a_h = 0$).

The total fodder-producing area, A_p (also in ha/pn), is

$$A_p = n_a \left(a_c M_c + a_s M_s + a_h M_h \right),$$

where we include the contribution for horses for completeness although it is neglected in any estimates. This land will be split between the various zones, following the seasonal farming model.

The dairy production produces the following milk yield Y_{mi} (in litre/pn/yr) and energy output E_m (in kcal/pn/yr):

$$Y_{mi} = n_a f_{dc} \left(\kappa_c a_c y_c + \kappa_s a_s y_s \right), \qquad E_m = n_a f_{dc} \left(\kappa_c a_c y_c e_{mc} + \kappa_s a_s y_s e_{ms} \right).$$

In terms of these quantities, we can re-derive our (prescribed) parameter $\epsilon_d = (E_a + E_m)/C$.

The per capita (hd/pn) and total (hd) animal populations are given by

$$n_c = n_a a_c, \quad n_s = n_a a_s, \quad n_{pi} = n_a a_{pi}, \quad n_h = n_a a_h,$$

and

$$N_c = N n_c, \quad N_s = N n_s, \quad N_{pi} = N n_{pi}, \quad N_h = N n_h.$$

For wild animals (adding chamoix and ibex to the red deer, roe deer and boar considered by Shukurov et al. 2015), we have

$$n_w = \frac{\epsilon_w C}{a_r m_r e_r + a_{ro} m_{ro} e_{ro} + a_b m_b e_b + a_{ch} m_{ch} e_{ch} + a_i m_i e_i},$$

$$E_w = n_w \left(a_r m_r e_r + a_{ro} m_{ro} e_{ro} + a_b m_b e_b + a_{ch} m_{ch} e_{ch} + a_i m_i e_i \right),$$

consistent with our (prescribed) parameter $\epsilon_w = E_w/C$. The annual hunting culls (per capita and total, respectively) are

$$n_r = n_w a_r, \quad n_{ro} = n_w a_{ro}, \quad n_b = n_w a_b, \quad n_{ch} = n_w a_{ch}, \quad n_i = n_w a_i,$$

and

$$N_r = N n_r, \quad N_{ro} = N n_{ro}, \quad N_b = N n_b, \quad N_{ch} = N n_{ch}, \quad N_i = N n_i.$$

For the crop productivity, we assume that manure is only gathered from cattle, and only when they are in the lowlands. Given our other parameters, the fraction of the crop fields which can be manured is

$$f_m = \left[1 - r_{mu} + \frac{C\mu(1+\delta_r)}{mn_c(1-\delta_{mid})(1-\gamma-\lambda)} \frac{\epsilon_g}{e_g Y_u}\right]^{-1}.$$

When potatoes are introduced as another staple crop (in Model 2), this is modified to

$$f_m = \left[1 - r_{mu} + \frac{C\mu(1+\delta_r)}{mn_c(1-\delta_{mid})(1-\gamma-\lambda)} \left(\frac{\epsilon_g}{e_g Y_u} + \frac{\epsilon_p}{e_p Y_{pu}}\right)\right]^{-1}.$$

Here we assume, for simplicity, that the use of manure increases the productivity of both grain and potatoes by the same ratio r_{mu}.

Using f_m, the various yields, areas and energy contributions can be calculated as

$$Y_m = r_{mu} Y_u, \quad Y = f_m Y_m + (1-f_m) Y_u, \quad Y_{pm} = r_{mu} Y_{pu}, \quad Y_{p,tot} = f_m Y_{pm} + (1-f_m) Y_{pu},$$

$$Y_g = (1-\gamma-\lambda)Y, \quad A_g = \frac{\epsilon_g C}{e_g Y_g}, \quad E_g = e_g Y_g A_g,$$

$$Y_p = (1-\gamma-\lambda)Y_{p,tot}, \quad A_{po} = \frac{\epsilon_p C}{e_p Y_p}, \quad E_p = e_p Y_p A_{po},$$

$$A_f = (1+\delta_f)(A_g + A_{po}).$$

The total crop areas (all in the lowlands, in ha) are then

$$A_l = NA_f, \quad A_{cereal} = NA_g, \quad A_{potato} = NA_{po}, \quad A_{fruit+veg} = A_l - A_{cereal} - A_{potato}.$$

The pasture and grazing areas (A_a in ha/pn and A_{graze} in ha) are calculated for highlands and midlands as

$$A_{a,high} = \frac{2\delta_{high}}{\delta_{mid} A_{a,trans}}, \quad A_{a,mid} = \left(1 - \frac{2\delta_{high}}{\delta_{mid}}\right) A_{a,trans},$$

$$A_{a,graze} = NA_{a,trans}, \quad A_{graze,high} = NA_{a,high}, \quad A_{graze,mid} = NA_{a,mid}.$$

Note that the high-altitude grazing land area calculated in this way is relatively small, compared to the area of high-altitude pastures actually available (and documented in the historical records). The size of the herds is effectively constrained by the areas available at other times of the year (including the requirement to store sufficient fodder for winter), rather than by the area of the high-altitude fields. As a result, the high-altitude areas calculated here must simply be regarded as a lower bound on the upland area actually used. In practice, the animals may well have been able to graze over considerably larger areas than required for summer subsistence (and this may have allowed for increased summer dairy yields, although this is not currently modelled).

Similarly, the haymaking and meadow areas (A_{hm} in ha/pn and A_{meadow} in ha, respectively) are calculated as

$$A_{hm} = \aleph A_p, \quad A_{hm,mid} = A_{a,mid} f_{hm,mid}, \quad A_{hm,low} = A_{hm} - A_{hm,mid},$$

$$A_{meadow} = NA_{hm}, \quad A_{meadow,mid} = NA_{hm,mid}, \quad A_{meadow,low} = NA_{hm,low}.$$

The parameter $f_{hm,\,mid}$ is explained below. Several aspects of these calculations deserve attention. As in the Neolithic model of Shukurov et al. (2015), the parameter $\aleph$ allows for a proportion $1 - \aleph$ of leafy fodder to be used in addition to the hay; thus, the area required for the latter is reduced. We also note that the mid-altitude areas are 'effective' areas corresponding to the area that would be required if used for 6 months (in terms of which our basic parameters such as M_c are set). Since the mid-altitude fields are only available for haymaking for half this time, the effective haymaking area is half the actual area; this is accounted for with the parameter $f_{hm,\,mid} = 1/2$. Further, the area of mid-altitude haymaking fields is directly connected to the area of mid-altitude pasture fields; these are the same fields whose use simply changes when the animals move to the high pastures in summer. The pasture requirements therefore dictate the mid-altitude land use in our model. As a result, the mid-altitude area cannot produce enough fodder for the winter requirements, and some additional meadow fields at low altitudes (near the villages) are required.

In reality, the low-altitude fields may well also have been used for pasture in early spring or late autumn (or indeed as pasture for non-transhumance animals – pigs and horses – which we currently treat differently or omit). And the 3-month/6-month land-use cycles which we use here could be refined (e.g. to require slightly less than 6 months of fodder over winter). Even though the present treatment is deliberately oversimplified, we obtain reasonable and plausible values for the areas of various regions, animal numbers and crop produces, demonstrating that the basic assumptions of the economy model are reasonable.

Labour Costs

Preliminary estimates have been made of the time devoted to harvesting the major crops in these models, such as reaping of cereals, cutting of grass for hay and, in Model 2, harvesting potatoes. These activities are of most interest as they correspond to the narrowest 'bottlenecks' in the model, where significant activity needs to be completed in a short span of time. If the models have been set up unrealistically, then the work required for these activities might not be realistically feasible. These calculations therefore act as a consistency check on our model.

In pre-mechanised agriculture, harvesting was an especially demanding task requiring significant community effort. In both models, we assume that 0.5 (50%)

of the population take part in these activities, including children who can participate in less arduous parts of the tasks. In both models, we assume the reaping of cereals (including the threshing) can be done at a rate of 30 m^2/person-hour; the cutting of meadow grass (which does not require threshing) can be done at a faster rate of 60 m^2/person-hour; harvesting potatoes is slower, progressing at 16 m^2/person-hour. The figures for cereal and grass are adapted from Shukurov et al. (2015) with allowance for more efficient, historical tools. The figure for potatoes is based on the data from Tripathi and Sah (2001) but reduced to consider only the harvesting labour rather than the total annual labour. This reduction employed a factor of roughly a third, this factor giving a suitable conversion for cereal crops.

In Model 1, the total cereal area requires 620 person-days to reap (based on a 10-hour working day). With 50% of the settlement population participating, this equates to 8.2 calendar days. The significant meadow area required to produce 6 months fodder poses a greater labour demand; however, the relevant area requires 1960 person-days to cut. Even with 50% participation, this translates to 26 calendar days.

For Model 2, the total cereal area requires 740 person-days to reap, which is only 4.9 calendar days for the larger settlement. The potato area requires 690 person-days or 4.6 calendar days. Again, cutting the meadows for fodder is the most significant constraint: this requires 2980 person-days, or 20 calendar days.

In addition to harvesting costs, the daily labour required for milking and the biannual labour required for shearing sheep are also calculated, again as consistency checks on our model.

For the milking calculations, we assume that it takes 15 minutes to milk each cow. In Model 1, we assume that four people are devoted to this task; in Model 2, we assume eight people. These numbers would correspond to two and four individuals, respectively, in each of two upland sites (in different valleys); for Model 2, only two of the individuals at each site would likely be adults. In Model 1, a total of 23 person-hours would be required daily; given the number of people involved, this is 5.7 clock hours. In Model 2, the milking requires 36 person-hours or 4.5 clock hours.

For the shearing calculations, we assume that it takes 10 minutes to shear each sheep. For both models, four individuals (skilled workers) would do this job. In Model 1, this would require 3.8 person-days or 0.94 calendar days per shearing season (i.e. twice per year). In Model 2, the numbers are 3.3 person-days or 0.82 calendar days.

Economic Production (Wool and Dairy)

In comparison with the Neolithic model of Shukurov et al. (2015), the more recent agricultural systems considered here are clearly more capable of producing surplus outputs, which can be converted to economic value in the more-developed economy of the time. In this work we concentrate on the surplus production of wool and dairy (cheese). The surplus is converted into monetary values using contemporary pricing records (Schmelzer 1972; Dietrich 1980) and thus into the amount of full-time labour (full-time equivalents: FTEs) – a more meaningful comparison for other costs – using contemporary labourer costs.

Some of the wool and dairy produce always remains for local consumption. We introduce model parameters f_{wc} and f_{dc} for the fractions of wool and dairy produce consumed locally. In Period 1, modelling a wool-producing economy (with a relatively large sheep herd, for the size of the settlement), $f_{wc} = 0.2$ is low but $f_{dc} = 0.9$ is high. In Period 2, modelling a dairy economy (with larger cattle herds), $f_{wc} = 0.8$ is higher, while $f_{dc} = 0.5$ is lower.

In terms of these parameters, the wool production Y_{wool} (kg/yr) is $Y_{wool} = (1 - f_{wc})y_w N_s$, where y_w is the production per year (kg/hd/yr). We adopt $y_w = 2.5$ kg/head, approximately 2/3 of the productivity of modern-day Alpine species (e.g. black-brown mountain sheep).

Adopting a historical value of 20 Kr/lb (Austro-Hungarian Krone/Habsburg civil pound) for the market price of wool for Model 1, the settlement in Model 1 produces 450 kg/year of surplus wool, earning an income of 16,000 Kr/year. Adopting then a historical value of 15 Kr/day for the wage of an adult male worker, this income is the equivalent of 1100 person-days, so that this local wool income is the equivalent of 4.3 FTE workers (based on 250 working days in the year), which the settlement could hire to work on other activities.

In Model 2, the corresponding costs are 24 Kr/lb for wool and 32 Kr/day for wages. The Model 2 settlement, producing 98 kg/yr of surplus wool, therefore earns 4200 Kr/year from wool sales: the equivalent of 130 person-days, or 0.53 FTE. (Recall that Model 2 is consuming most of its wool; most of its economic activity is based on dairy farming.)

Considering dairy, we assume that 0.1 kg of cheese can be produced from 1 litre of milk. Adopting then historical values for the market price of cheese – 3 Kr/lb for Model 1, and 9 Kr/lb for Model 2 – we can similarly estimate the profits of the dairy economy. In Model 1, the cheese produced (1400 kg/year) is worth 8300 Kr/yr or 2.2 FTE workers. In Model 2, more devoted to dairy, the 20,000 kg/yr of cheese produced generates an income of 330,000 Kr/yr, the equivalent of 41 FTE workers (using contemporary wages).

Discussion

Simulation results, presented in Table 2, clearly show how the change in farming strategies between the two periods led to a significant change in herd composition and in the overall number of domesticated animals. The first model simulated 91 cows, 227 sheep and 45 pigs in Ortisé and Menas for the mid-eighteenth century. The proportion of sheep and cows (5/2) fits well with an economic strategy primarily aimed at subsistence, including a nonspecialised focus on wool trading. Interestingly, the model estimated that the amount of wool and cheese produced and sold every year could contribute to the annual salary of 6.5 farmers, corresponding

Table 2 Model outputs

Category	Parameter	Name	M1	M2
Field areas	Cereal area (ha)	A_{cereal}	18.5	22.2
	Potato area (ha)	A_{potato}	–	11.1
	Fruit and vegetable area (ha)	$A_{fruit, veg}$	9.3	11.1
	Lowland grazing area (ha)	$A_{graze, low}$	–	–
	Mid-altitude grazing area (ha)	$A_{graze, mid}$	113.3	162.3
	High-altitude grazing area (ha)	$A_{graze, high}$	113.3	162.3
	Lowland meadow area (ha)	$A_{meadow, low}$	61.2	97.9
	Mid-altitude meadow area (ha)	$A_{meadow, mid}$	113.3	162.3
Arable yields	Cereal yield (kg/ha/yr)	Y	1265.2	1264.2
	Potato yield (kg/ha/yr)	$Y_{p,tot}$	–	10,535
	Fraction of fields manured	f_m	0.27	0.27
Domestic animal populations	Cattle (hd)	N_c	91	143
	Sheep/goats (hd)	N_s	227	196
	Pigs (hd)	N_{pi}	45	18
	Horses (hd)	N_h	–	–
Wild animal populations	Red deer (hd)	N_r	21	25
	Roe deer (hd)	N_{ro}	21	25
	Wild boar (hd)	N_b	11	13
	Chamois (hd)	N_{ch}	21	25
	Ibex (hd)	N_i	11	13
Labour costs	Cereal reaping time (day/yr)	T_{reap}	8.2	4.9
	Grass cutting time (day/yr)	T_{cut}	26.2	19.9
	Potato harvesting time (day/yr)	T_{pot}	–	4.6
	Sheep shearing time (day/season)	τ_{shear}	0.9	0.8
	Milking time (hr/day)	τ_{milk}	5.7	4.5
Wool economy	Wool production (kg/yr)	Y_{wool}	453	98
	Wool income (Kr/yr)	I_{wool}	16,200	4200
	Income labour equivalent (pn-day/yr)	$I_{wool, day}$	1080	131
	Income labour equivalent (FTE)	$I_{wool, FTE}$	4.3	0.5
Dairy economy	Cheese production (kg/yr)	Y_{cheese}	1550	20,300
	Cheese income (Kr/yr)	I_{cheese}	8320	326,000
	Income labour equivalent (pn-day/yr)	$I_{cheese, day}$	554	10,200
	Income labour equivalent (FTE)	$I_{cheese, FTE}$	2.2	40.7

Keys: *ha* hectare, *yr* year, *hd* head (animal), *pn* person, *hr* hour, *Kr* crowns (currency), *lb* Habsburg pound (weight)

to the 4.3% of the inhabitants. This scenario also suggests that each family in the villages (8 individuals per family = approximately 19 families) owned on average 5 cows, 12 sheep and 2 pigs. The few economic data available for the area and the period suggest that these figures are not unreasonable, although further investigations are needed to assess these preliminary outcomes. For the mid-nineteenth century, the model returns 143 cattle (+60% from the previous period), 196 sheep (−14%) and 18 pigs (−60%).

This scenario clearly fits with an alpine economy increasingly centred on dairy cattle and on the production and trading of dairy products. The decreasing economic value of wool in this period is clearly mirrored in the drop of sheep number; furthermore, an estimation of the revenues of wool trading shows that only 0.5 people in the village could hypothetically get a living by selling wool. On the other hand, the increasing importance of cheese can be deduced by the increase in the absolute number and proportion of cows and by the fact that 41 people (approx. 14% of the population) could get an annual salary by producing and selling cheese. Each of the 37 families estimated for this period had, on average, 4 cattle, 5 sheep and 0.5 pigs: the number of cattle per family decreased, and the number of sheep halved. These figures suggest the existence of "consortia" of families for the management of rural activities as documented in the area for more recent periods (Castiglioni 1976). On the other hand, the negligible importance of pigs is intriguing and would deserve further investigations.

The most significant and surprising result of the simulation is related to the assessment of land-use change in the two selected periods. The results of simulation for the mid-nineteenth century suggest that 44.4 ha around the two villages were occupied by cultivated land (1/4 rye, 1/4 barley, 1/4 potato, 1/4 fruit/vegetables) and 97.9 ha by meadows for fodder production. In the uplands, 162.3 ha were used for fodder and grazing animals during spring and autumn, and 162.3 ha were used as summer pastures. These results are remarkably similar to the actual land use of Ortisé and Menas described in the Habsburg cadastre of 1857. Here the surface of cultivated land was 36 ha, meadows near the villages 73 ha and meadows at high altitude 110 ha. The only value that deviates significantly from the simulation results is the surface of high-altitude pastureland (480 ha in the cadastre), but as pointed out above, the model simulates the lower bound of the area used. The validation of simulation results against the historical cadastre confirm that the parameters used for simulation and their estimated value describe quite accurately the economic strategies carried out in the study area during the transition to modern rural economy.

Another positive surprise of land-use simulation is the similarity between the first and the second model. The first model (mid-eighteenth century) is characterised by the absence of potato and by an animal husbandry strategy less intense and specialised than the following period. Nevertheless, the simulation for the eighteenth century produced a land-use scenario remarkably similar to the nineteenth century: 27.8 ha of the areas around the villages were used for cultivation and 61.2 ha for fodder; 113.3 ha of the mid-altitude meadows were used for foddering and 113.3 ha of the high-pastures for summer grazing. The increase in the surface used for agricultural and pastoral activities in the nineteenth century ranges between 40% (highland meadows and pastureland) and 60% (arable land and lowland meadows). This variation is rather moderate, considering the significant economic change occurred between the two periods and, most importantly, considering that population doubled between the eighteenth and the nineteenth centuries. A moderate increase of the different sectors used for farming purposes (from 315.6 ha to 466.9 ha, +47%), coupled with a significant increase in population (150 to 300, +100%), suggests that rural economy during the nineteenth century might have been more efficient than the subsistence strategies simulated for the eighteenth century. The even lower increase of the surface exploited at high altitude shows that despite the intensification in animal husbandry and the increase in large livestock (requiring larger grazing areas and more fodder for the winter), the exploitation of the vulnerable high-mountain environments was less intense and sustained a much larger population using roughly the same portion of the high-mountain environments. It can therefore be argued that, despite the a priori perception of premodern farming practices as more ecologically and economically sustainable than the modern ones, the transition from a subsistence-focused and sheep-based pastoral strategy to a dairy-focused and cattle-based strategy led to an increase in ecological sustainability. However, the specialisation of rural economy and the higher reliance on external sources of income might have weakened the resilience and risk mitigation strategies of local communities, thus increasing the vulnerability to macroeconomic fluctuations. The collapse of wool market at the end of the eighteenth century did not affect the local economy, as only a small portion of the inhabitants (4–5%) could live out of wool trading, whereas the number of people relying on cheese trading in the nineteenth century was three times higher, and so was their risk. This tendency towards an increasing specialisation and decreasing resilience continued and incremented during the early-twentieth century and might explain the collapse of local rural economy and the abandonment of mountain villages, widely documented in many sectors of the Alpine region (Rosenberg 1988). It must be acknowledged that the money acquired from cheese trading could have been reinvested in activities aimed at improving the productivity of farming activities (e.g. farming tools, new breeds, etc.). However, if this mechanism contributed to increasing the short-term yields and gains, it did not compensate the decreasing ability of local economy to recover from potential disturbances due to macroeconomic trends.

Conclusions

As pointed out before, the models presented in this publication represent just the first step in the development of more sophisticated dynamic models which will be able to address the complexity of socioecological dynamics in mountain environments. But despite being preliminary, they produced promising and interesting results, whose implications and methodological significance will be rapidly discussed in this final section.

The assessment of the unexpectedly higher economic efficiency of the nineteenth-century intensive farming shows how useful computer simulations can be to challenge general perceptions. It can be argued that the approximation of some of the parameters and the static nature of the model prevents a realistic evaluation of the reliability of these results. But the remarkable similarity of the simulated land use with the 1850s cadastre suggests that the set of parameters selected from ethnographic and ethnohistorical investigation are good proxies for delineating the subsistence economy of an Alpine small-scale community. This in turn shows the importance of validating the model using the same ethnographic context where the model's parameters have been defined, in order to assess the accuracy of the decision-making rules and the reliability of the model's outcomes. Furthermore, this assessment phase might help to calibrate the model according to the characteristics of the wider socio-economic context. For example, it is clear that the efficiency of economic strategy simulated for the nineteenth century is justifiable only in a market economy, where the revenue of cheese and butter trading lower the pressure of increasing population on local resources. If we decrease the impact of dairy products demand in the model (e.g. by assuming that the 80% of cheese/butter is consumed locally), the ecological sustainability suggested before is expected to drop, along with local economic resilience.

These observations suggest that the ethnoarchaeological approach improves the accuracy of hypothetical scenarios for archaeological simulation, particularly in mountain socio-economical systems characterised by the coexistence of permanent and seasonal farming. The use of decision-making rules inferred from ethnographic contexts and the calibration of simulation scenarios using the data available for the same contexts are more solid than simply using archaeological datasets to test simulation results, as the available dataset might be significantly biased or so poor to prevent a reliable estimation of the model performance. It is worth pointing out that we do not question the comparison of simulation results with archaeological data to assess the explanatory value of the models, since the ultimate purpose is the identification of realistic socioecological scenarios for the past. We just argue that ethnoarchaeological validation and calibration should be an important additional step of dynamic system modelling, to be carried out before archaeological model validation. The preliminary results presented in this paper suggest the solidity of this approach, which will be further developed and applied to archaeological contexts, to test its heuristic potential beyond methodological coherence and significance.

Acknowledgements Authors are indebted to Professors Andrea Bonoldi, Marco Bellabarba, Emanuele Curzel and Giuseppe Albertoni (University of Trento) and Dr Italo Franceschini (San Bernardino Library, Trento) for sharing first-hand historic and economic data and for providing useful suggestions.

Research in Val Molinac and Val Poré is conducted in the context of the ALPES project. The project is codirected by D.E.A. and F.C., is carried out with the collaboration of the *Ufficio Beni Archeologici, Soprintendenza per i Beni Culturali* of the autonomous province of Trento and has been funded by the *Dipartimento di Lettere e Filosofia* of the University of Trento, the GAL (*Gruppo Azione Locale*) of Val di Sole and the *Terre Alte* programme of CAI (*Club Alpino Italiano*), with the logistical support of the municipality of Mezzana.

References

Angelucci, D. E., & Carrer, F., (Eds.) (2015). *Paesaggi pastorali d'alta quota in Val di Sole (Trento). Le ricerche del progetto ALPES – 2010-2014*. Dipartimento di Lettere e Filosofia, Università di Trento, Trento.
Angelucci, D. E., Carrer, F., & Cavulli, F. (2014). Shaping a periglacial land into a pastoral landscape: A case study from Val di Sole (Trento, Italy). *Post-Classical Archaeologies, 4*, 157–180.

Angelucci, D. E., Carrer, F., & Pedrotti, A. (2017). Due nuove datazioni dell'età del Bronzo da un sito d'alta quota in Val Poré (Val di Sole). *Archeologia delle Alpi, 2016*(7), 154–156.

Balbo, A., Baggethun-Gomez, E., Salpeteur, M., Puy, A., Biagetti, S., & Scheffran, J. (2016). Resilience of small-scale societies: A view from drylands. *Ecology and Society*. URL: http://www.ecologyandsociety.org/volXX/issYY/artZZ/.

Barceló, J. A., et al. (2015). Simulating Patagonian territoriality in prehistory: Space, frontiers and networks among hunter-gatherers. In G. Wurzer, K. Kowarik, & H. Reschreiter (Eds.), *Agent-based modelling and simulation in archaeology* (pp. 217–256). Cham: Springer.

Barker, G., & Grant, A. (1991). Ancient and modern pastoralism in Central Italy: An interdisciplinary study. *Papers of the British School at Rome, 59*, 15–88.

Barton, C. M., Ullah, I. I., & Bergin, S. (2010). Land use, water and Mediterranean landscapes: Modelling long-term dynamics of complex socio-ecological systems. *Philosophical Transactions of the Royal Society A: Mathematical, Physical and Engineering Sciences, 368*(1931), 5275–5297.

Battisti, C. (1904). Noterelle statistiche sul bestiame da pascolo, le malghe, le latterie e l'industria dei latticini nel Trentino. *Tridentum, 4*, 159–173.

Bennett, E. M., Cramer, W., Begossi, A., Cundill, G., DÍaz, S., Egoh, B. N., Geijzendorffer, I. R., Krug, C. B., Lavorel, S., Lazos, E., Lebel, L., MartÍn-López, B., Meyfroidt, P., Mooney, H. A., Nel, J. L., Pascual, U., Payet, K., Harguindeguy, P. N., Peterson, G. D., Prieur-Richard, A.-H., Reyers, B., Roebeling, P., Seppelt, R., Solan, M., Tschakert, P., Tscharntke, T., Turner, B. L., II, Verburg, P. H., Viglizzo, E. F., White, P. C. L., & Woodward, G. (2015). Linking biodiversity, ecosystem services, and human well-being: Three challenges for designing research for sustainability. *Current Opinion in Environmental Sustainability, 14*, 76–85.

Biagetti, S. (2014). *Ethnoarchaeology of the Kel Tadrart Tuareg. Pastoralism and resilience in Central Sahara*. Heidelberg: Springer.

Biagetti, S., Alcaina-Maeos, J., & Crema, E. (2016). A matter of ephemerality. The study of Kel Tadrar Tuareg campsites via quantitative spatial analysis. *Ecology and Society, 21*(1), 42. https://doi.org/10.5751/ES-08202-210142.

Boserup, E. (1965). *The conditions of agricultural growth*. London: Allen & Unwin.

Bourgarit, D., Rostan, P., Burger, E., Carozza, L., Mille, B., & Artioli, G. (2008). The beginning of copper mass production in the western Alps: The Saint-Veran mining area reconsidered. *Historical Metallurgy, 42*(1), 1–11.

Brisset, E., Guiter, F., Miramont, C., Troussier, T., Sabatier, P., Poher, Y., Cartier, R., Arnaud, F., Malet, E., & Anthony, E. J. (2017). The overlooked human influence in historic and prehistoric floods in the European Alps. *Geology*. https://doi.org/10.1130/G38498.1.

Briz i Godino, I., et al. (2014). Social cooperation and resource management dynamics among late hunter-fisher-gatherer societies in Tierra del Fuego (South America). *Journal of Archaeological Method and Theory, 21*, 343–363.

Carrer, F. (2013). An ethnoarchaeological inductive model for predicting archaeological site location: A case study of pastoral settlement patterns in the Val di Fiemme and Val di Sole (Trentino, Italian Alps). *Journal of Anthropological Archaeology, 32*, 54–62.

Carrer, F. (2015). Herding strategies, dairy economy and seasonal sites in the Southern Alps: Ethnoarchaeological inferences and archaeological implications. *Journal of Mediterranean Archaeology, 28*(1), 3–22.

Carrer, F. (2017). Interpreting intra-site spatial patterns in seasonal contexts: An ethnoarchaeological case study from the Western Alps. *Journal of Archaeological Method and Theory, 24*, 303–327.

Carrer, F., & Angelucci, D. E. (2017). Continuity and discontinuity in the history of upland pastoral landscapes: The case study of Val Molinac and Val Poré (Val di Sole, Trentino, Eastern Italian Alps). *Landscape Research, 43*, 862. https://doi.org/10.1080/01426397.2017.1390078.

Carrer, F., Mocci, F., & Walsh, K. (2015). Etnoarcheologia dei paesaggi alpine di alta quota nelle Alpi occidentali: un bilancio preliminare. *IL Capitale Culturale, 12*, 621–635.

Carrer, F., Colonese, A. C., Lucquin, A., Petersen Guedes, E., Thompson, A., Walsh, K., Reitmaier, T., & Craig, O. E. (2016). Chemical analysis of pottery demonstrates prehistoric origin for high-altitude alpine dairying. *PLoS One, 11*(4), e0151442. https://doi.org/10.1371/journal. pone.0151442.

Castiglioni, G. B. (1976). Ortisé, un piccolo centro rurale della Val di Sole. *Aspetti geografici del Trentino-Alto Adige occidentale, 34. Escursione geografica interuniversitaria (settembre 1974)* (pp. 179–192). Universita di Padova, Istituto di Geografia.

Catalan, J., Ninot, J. M., & Aniz, M. M. (2017). The high mountain conservation in a changing world. In J. Catalan, et al. (Eds.), *High mountain conservation in a changing worlds* (pp. 3–36). Cham (Switzerland): Springer.

Chang, C., & Tourtellotte, A. (1993). The Ethnoarchaeological survey of pastoral transhumance sites. *Journal of Field Archaeology, 20*(3), 249–264.

Christie, N., Beavitt, P., Gisbert Santoja, J., Gil Senís, V., & Seguí, J. (2007). Peopling the recent past in the Serra de L'Altmirant: Shepherds and farmers at the margins. *International Journal of Historical Archaeology, 11*, 304–321.

Cole, J. W., & Wolf, E. R. (1974). *The hidden frontier: Ecology and ethnicity in an Alpine Valley.* New York: Academic Press.

Cribb, R. (1984). Computer simulation of herding systems as an interpretative and heuristic device in the study of kill-off strategies. In J. Clutton-Brock, & C. Grigson (Eds.), *Animals and archaeology: 3. Early Herders and Their Flocks.* (BAR International Series 202, pp. 161–171). Oxford: Archaeopress.

Cunningham, J. J. (2009). Ethnoarchaeology beyond correlates. *Ethnoarchaeology, 1*(2), 115–136.

Danielisová, A., Olševičová, K., Cimler, R., & Machálek, T. (2015). Understanding the Iron age economy: Sustainability of agricultural practices under stable population growth. In G. Wurzer, K. Kowarik, & H. Reschreiter (Eds.), *Agent-based modeling and simulation in archaeology* (pp. 183–215). Cham: Springer.

David, N., & Kramer, C. (2001). *Ethnoarchaeology in action.* Cambridge: Cambridge University Press.

Dearing, J., Braimoh, A. K., Reenberg, A., Turner, B. L., & van der Leeuw, S. (2010). Complex land systems: The need for long time perspectives to assess their future. *Ecology and Society, 15*(4). URL: http://www.ecologyandsociety.org/vol15/iss4/art21/.

Dell'Amore, F., Carrer, F., & Angelucci, D. E. (2017). Reperti archeologici dalla Val Molinac e dalla Val Poré (Val di Sole, Trento, Italia). In L. Guerri, & N. Pedergnana (Eds.), *Archeologia e Cultura in Val di Sole. Ricerche Contesti Prospettive, Atti del Convegno – Molino Ruatti 10-11/09/2016* (pp. 131–143). Associazione Molino Ruatti, Rabbi.

Dietrich, W. (1980). *Geschichte der Preise und Löhne in Stift Stams von 1532 bis 1806.* (PhD Thesis). Leopold Franzens University, Innsbruck (Austria).

Dossche, R., Rogge, E., & Van Eetvelde, V. (2016). Detecting people's and landscape's identity in a changing mountain landscape. An example from the northern Apennines. *Landscape Research, 41*(8), 934–949.

Fernández-Giménez, M. E., & Fillat Estaque, F. (2012). Pyrenean pastoralists' ecological knowledge: Documentation and application to natural resource management and adaptation. *Human Ecology, 40*(2), 1–14.

Filatova, T., Verburg, P. H., Parker, D. C., & Stannard, C. A. (2013). Spatial agent-based models for socio-ecological systems: Challenges and prospects. *Environmental Modelling & Software, 45*, 1–7.

Fontana, F., & Visentin, D. (2016). Between the Venetian Alps and the Emilian Apennines (Northern Italy): Highland vs. lowland occupation in the early Mesolithic. *Quaternary International, 423*, 266–278.

Garde, L., Dimanche, M., & Lasseur, J. (2014). Permanence and changes in pastoral farming in the Southern Alps. *Journal of Alpine Research, 102*(2), 2–12. https://doi.org/10.4000/rga.2416.

Gleirscher, P. (2010). Hochweidenutzung oder Almwirtschaft? Alte und neue Überlegungen zur Interpretation urgeschichtlicher und römerzeitlicher Fundstellen in den Ostalpen. In F. Mandl, & H. Stadler (Eds.), *Archäologie in den Alpen. Alltag und Kult* (pp. 43–62). Forschungsberichte der ANISA. ANISA, Haus i.E. (Austria).

Gregg, S. A. (1988). *Foragers and farmers: Population interaction and agricultural expansion in prehistoric Europe*. Chicago: University of Chicago Press.

Halstead, P. (1996). Pastoralism or household herding? Problems of scale and specialization in early Greek animal husbandry. In K. D. Thomas (Ed.), *Zooarchaeology: New Approaches and Theory. World Archaeology, 28*(1), 20–42.

Lake, M. W. (2014). Trends in archaeological simulation. *Journal of Archaeological Method and Theory, 21*, 258–287.

Lancelotti, C., Negre, J., Alcaina Mateos, J., & Carrer, F. (2017). Intra-site spatial analysis in ethnoarchaeology. *Environmental Archaeology, 22*, 354. https://doi.org/10.1080/14614103.2017.1299908.

Le Couédic, M. (2012). Modéliser les pratiques pastorales d'altitude dans la longue durée. *Cybergeo: European Journal of Geography*. https://doi.org/10.4000/cybergeo.25123.

Mathieu, J. (2009). *History of the Alps 1500-1900. Environment, development, and society*. Morgantown: Virginia University Press.

Mientjes, A. C. (2004). Modern pastoral landscapes on the island of Sardinia (Italy). Recent pastoral practices in local versus macro-economic and macro-political contexts. *Archaeological Dialogues, 10*(2), 161–190.

Netting, R. M. (1981). *Balancing on an Alp: Ecological change and continuity in a Swiss mountain community*. Cambridge: Cambridge University Press.

Orme, B. (1981). *Anthropology for archaeologists: An introduction*. London: Gerald Duckworth.

Premo, L. S. (2007). Exploratory agent-based models: Towards an experimental ethnoarchaeology. In J. T. Clark, & E. M. Hagemeister (Eds.), *Digital discovery: Exploring new frontiers in human heritage. CAA 2006. Computer applications and quantitative methods in archaeology* (pp. 29–36). Archeolingua, Budapest.

Previtali, F. (2011). Mountain Anthroposcapes, the case of the Italian Alps. In S. Kapur & H. Eswaran (Eds.), *Sustainable land management: Learning from the past for the future* (pp. 143–161). Heidelberg: Springer.

Reitmaier, T., Doppler, T., Pike, A. W., Deschler-Erb, S., Hajdas, I., Walser, C., & Gerling, C. (2017). Alpine cattle management during the Bronze age at Ramosch- Mottata, Switzerland. *Quaternary International, 484*, 19. https://doi.org/10.1016/j.quaint.2017.02.007.

Rescia, A. J., Pons, A., Pons, A., Lomba, I., Estaban, C., & Dover, J. W. (2008). Reformulating the social-ecological system in a cultural rural mountain landscape in the Picos de Europa region (northern Spain). *Landscape and Urban Planning, 88*, 23–33.

Rondelli, B., Lancelotti, C., Madella, M., Pecci, A., Balbo, A., Ruiz Pérez, J., Inserra, F., Gadekar, C., Cau Ontiveros, M. Á., & Ajithprasad, P. (2014). Anthropic activity markers and spatial variability: An ethnoarchaeological experiment in a domestic unit of Northern Gujarat (India). *Journal of Archaeological Science, 41*, 482–492.

Rosenberg, H. G. (1988). *A negotiated world: Three centuries of change in a French Alpine Community*. Toronto: University of Toronto Press.

Schmelzer M. (1972). *Geschichte der Preise und Lohne in Rattenberg vom Ende des 15. bis in die 2. Halfte des 19. Jahrhnderts* (PhD Thesis). Leopold Franzens University, Innsbruck (Austria).

Schmidl, A., & Oeggl, K. (2005). Subsistence strategies of two Bronze age hill-top settlements in the eastern Alps-Friage/Bartholomäberg (Voralberg, Austria) and Ganglegg/Schluderns (South Tyrol, Italy). *Vegetation History & Archaeobotany, 14*, 303–312.

Shukurov, A., Sarson, G., Videiko, M., Henderson, K., Shiel, R., Dolukhanov, P., & Pashkevich, G. (2015). Productivity of premodern agriculture in the Cucuteni-Trypillia area. *Human Biology, 87*(3), 235–283. arXiv:1505.05121 [q-bio.PE].

Skibo, J. M. (2009). Archaeological theory and snake-oil peddling. The role of ethnoarchaeology in archaeology. *Ethnoarchaeology, 1*(1), 27–56.

Surovell, T. A., & O'Brien, M. (2016). Mobility at the scale of meters. *Evolutionary Anthropology, 25*(3), 142–152.

Tripathi, R. S., & Sah, V. K. (2001). Material and energy flows in high-hill, mid-hill and valley farming systems of Garhwal Himalaya. *Agriculture, Ecosystems and Environment, 86*, 75–91.

Varanini, G. M. (2004). L'economia. Aspetti e problemi (XIII-XV secolo). In A. Castagnetti, & G. M. Varanini (Eds.), *Storia del Trentino, III. L'età medievale* (pp. 461–516). Il Mulino, Bologna.

Viazzo, P. P. (1989). *Upland communities: Environment, population and social structure in the Alps since the sixteenth century*. Cambridge: Cambridge University Press.

Viazzo, P. P., & Woolf, S., (Eds.) (2001). *L'alpeggio e il mercato*. La Ricerca Folklorica 43.

Walsh, K., Court-Picon, M., de Beaulieu, J. L., Guiter, F., Mocci, F., Richer, S., Sinet, R., Talon, B., & Tzortzis, S. (2014). A historical ecology of the Ecrins (Southern French Alps): Archaeology and palaeoecology of the Mesolithic to the Medieval period. *Quaternary International, 353*, 52–73.

Wylie, A. (1982). An analogy by any other name is just as analogical: A commentary on the Gould-Watson dialogue. *Journal of Anthropological Archaeology, 1*, 382–401.

Zaninelli, S. (1979). *Una agricoltura di montagna nell'Ottocento: il Trentino*. Società di Studi Trentini di Scienze Storiche, Trento.

Trowels, Processors and Misunderstandings: Concluding Thoughts

Mehdi Saqalli and Marc Vander Linden

Supporting the Holistic Hope of Archaeology

Archaeology, under its many theoretical variants, has always had a complex relationship with its self-acknowledged role as the gatekeeper of the past intricacy of all facets of human life over the very *longue durée*. Having to deal with a record first and foremost characterised by patchiness and unevenness is to some extent counter-balanced by extraordinary access to an exciting diversity of sources, reflected by the countless methodological and technical refinements of the last few decades. We know that our window unto the past will always remain clouded in much haze, but we are constantly nudging it a bit more open. As a result, and perhaps more so than any other social sciences, most archaeologists – explicitly or not – seem to subscribe to a Zeitgeist dominated by the idea that their discipline's goal is to explore the complexity of cultural and social human life in its entirety while at the same time avoiding the meta-synthesis "original sin", more or less the same positioning as anthropologists for present-time social science.

Well in line with the wider intellectual ethos of their time, foundational texts of what came to be known as processual archaeology are dominated by one of the various versions of system theory (e.g. Doran 1970, Salmon 1978). This trend is most noticeable in David Clarke's contribution with his conceptualisation of several specialised subsystems all locked in many interaction and feedback loops (Clarke 1968). Regardless of the merits and pitfalls of this approach, it is obvious that its underlying rationale was to provide a – overwhelmingly – formal total view of past

---

M. Saqalli (✉)
UMR CNRS 5602 GEODE Géographie de l'Environnement, Maison de la Recherche, Université Toulouse 2 Jean Jaurès, Toulouse, France
e-mail: mehdi.saqalli@univ-tlse2.fr

M. Vander Linden
Department of Archaeology, University of Cambridge, CB2 3DZ, Cambridge, UK

human lifeways, a holistic ambition which permeated Clarke's entire oeuvre covering archaeological terminology and even fieldwork strategy (Evans et al. 2006). The holistic aspiration of archaeology is perhaps even more obvious in the following post-processual archaeology, with its constant obsession of all facets of human life permeating one another (e.g. Shanks and Tilley 1987; Hodder 1991). All in all, despite the apparent turbulence of theoretical agendas, a strong case can be made for archaeology constant interest towards all-encompassing approaches, perhaps as reaction to the sketchiness of its empirical foundations.

These two contrasted examples illustrate archaeology's hesitation between reassuring empirism and holistic pretence. Our purpose is not to challenge this pretence but rather to methodologically and practically support it. This goal indeed allows us to conceptualise in a different way computational modelling. As stated on many occasions in the previous pages, computational models, and especially agent-based models, enable the formal exploration of the interaction between a wide range of parameters. As pointed out in the introduction, the tension between "modelling" and "other" archaeologists does therefore not rest in their objectives but in the ways and intellectual decisions taken to achieve them. Modelling implies assuming common rules among archaeological sites: a modeller supposes that there are an equivalent set of rules among archaeological sites, even if these rules are adapted to local conditions, while non-modelling archaeologist acknowledges the necessity of a system ruling locally the different elements (architecture, practices, environment, etc.) but keeping variability at the forefront by not presupposing shared common rules. Especially crucial here is the notion, mentioned on previous occasions, of simplicity, or at least the simplification linked to modelling. While a recurring argument against simulation approaches, the modelling community sees it as an imperative requirement. The drawback of such self-imposed formality is however to expose in crude light the assumptions linking the various parameters, thus making such models easily prone to criticisms. It is fair to say that such criticisms, in many instances, actually make a relevant point, in the sense that the modellers perhaps pay too much attention upon the elucidation of the formal relationships between parameters, seen as a way to avoid deterministic accounts whereby a single factor trumps all others. Another implication is the fact that indeed and especially in the eyes of outsiders, modellers may overlook some considerations in their selection of parameters. This last point is perhaps made most salient by Roux about the choice of morphometric, rather than technical traits of cultural transmission, and also lies at the core of the methodological designs exposed by Saqalli and colleagues, Le Néchet and colleagues, as well as Carrer and colleagues.

A Call for Interdisciplinary-Formalized Archaeology Modelling

There is no single, easy answer to the question of how to best fix parameters. Two remarks however come to mind in view of the present contributions:

Firstly, a successful strategy for qualitative assessments in computational modelling lies in the oft-quoted interdisciplinary nature of the enterprise. This integration of specialists from different fields can occur, as shown by many contributions, at various stages of the process, either towards the very beginning of the model design or later through inclusion/reassessment of parameters following a first run of simulations. Although the addition and constant re-evaluation of parameters may come at a computational prize (i.e. simulations becoming increasingly demanding and thus long to run), the mathematical flexibility of agent-based models provides the required methodological – relative – ease for interdisciplinary work. Beyond these technicalities, it is essential to recognise that interdisciplinarity does not merely translate into the ever addition of parameters but also comes with what Saqalli and colleagues call sacrifice on the behalf of all partners. For the integration of various parameters to be effective, decisions, including simplifications, have to be made by all contributors. This notion of sacrifice is the key, both for its methodological implications and for its practical, social dimension: any successful interdisciplinary dialogue – and perhaps even so in the case of computational modelling – rests upon hard-fought equality among all participants. Archaeologists, ecologists and computer scientists, regardless of each one expertise, are considered as equal, with the same voice in the process. Here lies the fact that eventually, gentlemen's agreement and a solid argumentation may become the sole legitimacies for selecting and discriminating parameters and factors among an interdisciplinary community. This may sound both awkward and obvious but must be stressed in practical terms.

Secondly, as archaeological data provide an essential resource in the validation of models, at least parts of them must be designated for this validation step and thereby separated from the designing stage of the models. At best, inference from archaeological models can be of use in delineating a parameter space such as a range of values to be explored through simulations, as, for instance, in Approximate Bayesian Computation approaches. This situation apart, archaeological data may inform decisions regarding the selection of parameters, but cannot play a direct role in the qualitative assessment of these parameters. This point may prove contentious for many archaeologists, for whom data often constitute an ultimate reference but is imperative to avoid circular arguments.

Main Working Institutional Obstacles Beyond the Interdisciplinarity Mantra

This ideal interdisciplinary community situation is rarely met: as stated in our introduction, the rationale of this collection of papers stemmed from the recognition that, although computational modelling was arguably more popular than ever in archaeological circles, it still remains a niche activity. We hypothesize that this state of affairs indicates that the positioning of computational modelling within archaeology remains unclear and, more worryingly, that the practical integration of modelling concepts and methods with the archaeological toolkit and archaeological data is not working. While one may consider this situation as the outcome of a series of misunderstandings between modelling and field archaeologists regarding the notions of simplicity, approximation and eventually truth in the epistemological roots of archaeology, we position these misunderstandings into a more practical debate, where global syntheses remain underrated.

Beyond the epistemological limits, the restraining role of several factors ought not to be overlooked, including:

- The discipline-based organisation of numerous journals was an initial constraint, though one admittedly positively and rapidly changing.
- As an opposite, one may point out that institutional evaluation, in institutions keeping tenured positions and during application processes, remains segregated between disciplines and even regressing. Even more, the repeated call for interdisciplinarity looks much more the same like in other disciplines where interdisciplinarity becomes pluridisciplinarity, i.e. the juxtaposition of disciplines with no interaction.
- Although the situation varies enormously across countries, in several instances, funding remains allocated on a discipline base. As a consequence, modelling work packages are often portions of projects but rarely their main goals. Beyond the financial constraints it induces, it means that specialists are focused on delivering their own work packages and not necessarily ready to invest time in the modelling procedure, let alone to sacrifice the complexity of their topic for what remains a side objective.
- This last point is reinforced by an often misperception encountered by some of us, to perceive computational modelling as some sort of "subcontracting", merely added to provide "quantitative shine" or "scientific kudos" to a research proposal, for instance. It is noteworthy that this situation is hardly new and that similar concerns have been repeatedly voiced over the years when various archaeological techniques were developed and then used and abused, before a successful integration became the norm. This had happened as well in other disciplines such as geography and GIS.

As a humoristic conclusion, one may then, borrowing the Kübler-Ross five stages of grief, propose the five steps of modelling adoption in archaeology:

1. Enthusiasm: modelling is seen as panacea, finding the missing data, solving the flaws in hypotheses, providing beautiful illustrations and maps for articles and generating more funds from funding agencies.
2. Denial: modelling is acknowledged to be highly demanding and time-consuming, implying negotiations between scientists and oversimplification.
3. Reject and anger: modelling may even be a risk of overpowering field archaeologists within the discipline due to this "quantitative shining scientifically trendy kudos".
4. Bargaining: modelling is only a support for improving the process formalizing, combining and deducing archaeological reasoning. Once used properly, it is a powerful weapon for facilitating research, both facing funding agencies and reluctant institutes.

5. Cold acceptance: multidisciplinary project teams slowly self-impose themselves, thanks to non-niche funding practices.

We hypothesize that we all pass the phase 1 reaching the phase 2 and even the phase 3 quite abruptly, and we are observing in most cases goes-and-returns between phases 2 and 3 in one hand and phase 4 on the other hand with the latter being obviously much more rewarding.

Qualitative Factors in Modelling

Struggling with Qualitative Factors

In 2000, the modeller and sociologist Edmund Chattoe-Brown asked, during the 24th International Conference of Agricultural Economists (IAAE) "Why is building Multi-Agent Models of social systems so difficult?" (Chattoe 2000). Nearly two decades later, the chapters presented here did not shy away from the complexity encountered in building models integrating qualitative factors. Some appear more at first sight to be relative failures, such as Saqalli and colleagues, Barceló and colleagues and Bentley and O'Brien, while Le Néchet and colleagues discuss the obstacles encountered during the modelling process.

These apparent failures and struggles underline the inherent difficulty in tackling qualitative socio-anthropological factors independently from time (this difficulty concerns simulations of both past and present-time societies) and tools (the difficulty can be felt with or without computational modelling) and stress the necessity to address the fundamental issue of the adequacy between archaeological data and the requirements of modelling and statistical approaches.

As a result, since qualitative factors have by definition, and regardless of modelling, to be considered in any future interpretation of past societies, these difficulties may well be simply due to the fact that, after all, we are still in the early stages of formalizing qualitative factors. However, such integration implies solving or at least improving the methodological formalisation of qualitative factors. This challenge is essential in both modelling, and, beyond, in any interpretative system as, in all instances, priorities regarding the inclusion or exclusion of factors should be made explicit.

Balancing Between Approaches for KISS and KIDS for Modelling RSES

Therefore, although qualitative factors may in many cases be partly dependent upon biophysical constraints, ethnography and anthropology demonstrate the huge variety of social combinations possible for each biophysical environment. Given that, by definition, it remains impossible, if only from a practical point of view, to offer any total description of any society and its biophysical environments, the selection of factors to be considered remains a crucial, though perhaps somehow overlooked, intellectual task. From this point of view, and going back to the KIDS vs. KISS opposition (Edmonds and Moss 2005; Moss and Edmonds 2005a, b), one should clarify when it is relevant to:

- Work on a society as a whole implies an empirical "KIDS" approach (Keep It Descriptive, Stupid"): it sometimes leads to an inflation of factors included and results in fuzzy results. Such models often require the integration of parameters for which direct analogues are either lacking or extremely remote, thus lowering the quality requirements of their integration. As a result, KIDS can never be complete, meaning they cannot cover and model all aspects of past social life. A selection of main issues is to be assessed.
- Focus on one social phenomenon, implying a possibly more operative and straightforward "KISS" approach (Keep It Simple, Stupid"), while it possibly leads to apparently more precise results, this approach requires explicit justification of the choice and the relevance of the limited sets of parameters exposed, unless simulating dynamics that never occurred in reality, overpowered by neglected and non-simulated factors. As a result, KISS can never be exact, meaning they can never completely explain the priority given to the chosen dynamics.

Combining these elements is then necessary, and we support the acknowledgement that modelling the past is always a balance between KIDS and KISS extreme poles, as a medium way for simulating rural socio-ecological systems (RSES). This balance should be always justified according to the importance of the selected dynamic to simulate.

Beyond Socioecological Systems

Indeed, regarding RSES, biophysical parameters are positive constraints that both increase local conditionalities, reducing the dispersal of social hypotheses but also provide testing data for validating purposes, giving more confidence to modelling approaches. We then here question the challenge of modelling factors and societies that are poorly conditioned by such biophysical factors, such as norm rules and customs, non-farming social classes and urban societies. That would be the next challenge for modelling archaeologists. A possible way is to focus on ceteris paribus situations where other factors apart from the studied social issues remain or can be assumed as stable. It means simulating over long but very stable periods of time, over short periods of time or even over very short periods with only the studied dynamic.

As repeatedly said by Charles Darwin, "one cannot be a good observer without being at the same time an active theorist". A good archaeologist has to be also a conceptualist, and modelling should then be seen as the support for this demarche balancing between field and formalisation. One may then consider that the future of archaeology will increasingly include modelling tools in the many historical cases where biophysical constraints failed to be sufficient enough for explaining past population dynamics.

References

Chattoe, E. (2000). *Why is building multi-agent models of social systems so difficult? A case study of innovation diffusion*. 24th International Conference of Agricultural Economists (IAAE).
Clarke, D. (1968). *Analytical archaeology*. London, UK: Methuen.
Doran, J. (1970). Systems theory, computer simulations and archaeology. *World Archaeology, 1*(3), 289–298.
Edmonds, B., & Moss, S. (2005). From KISS to KIDS: An "anti-simplistic" modelling approach. *Lecture Notes in Artificial Intelligence, 34*, 130–144.
Evans, C., Edmonds, M., Boreham, S., Evans, J., Jones, G., Knight, M., & Legge, T. (2006). 'Total archaeology' and model landscapes: Excavation of the great Wilbraham causewayed enclosure, Cambridgeshire, 1975–76. *Proceedings of the Prehistoric Society, 72*, 113–162. https://doi.org/10.1017/S0079497X00000803.
Hodder, I. (1991). *Reading the past: Current approaches to interpretation in archaeology*. Cambridge, UK: Cambridge University Press.
Moss, S., & Edmonds, B. (2005a). Towards good social science. *Journal of Artificial Societies and Social Simulation, 8*, 13.
Moss, S., & Edmonds, B. (2005b). Sociology and simulation: Statistical and qualitative cross-validation. *American Journal of Sociology, 110*, 1095–1131.
Salmon, M. H. (1978). What can systems theory do for archaeology? *American Antiquity, 43*(2), 174–183.
Shanks, M., & Tilley, C. (1987). *Re-constructing archaeology: Theory and practice*. London, UK: Routledge.

Index

A
Accuracy of Variables then Inventory of Data (AVID), 27, 28
Agent-based models, 1, 2, 17, 218
　distortions and wide variations, 35
　spatial reconstruction of interactions, RSES, 23, 24
Agents, 17
　ethnic group, 57
　mobile hunter-gatherers, 68
　restriction of interaction, 59
　social, 55
　surplus, 70
Anteriority, 33
Anthropological rules, 32
Archaeology
　cladistic analysis, 67
　computational modelling, 1, 2
　cultural analysis, 66
　definition, 4
　ethnicity (*see* Ethnicity)
Axelrod model, 71

B
Bantu expansion, 109
　agent-based models, 113
　Bantu farmers and FF, 117
　hypothesis, 132
　interactions, 116, 117
　MAS models, 119
　peopling of empty land, 114, 115
　Sub-Sahara Africa, 113
Bantu farmers, 124

Bantu migrations
　demographic expansion (*see* Bantu expansion)
　relationship, forest foragers, 112
　scenarios, 110, 111

C
Chaîne opératoire approach, 9
Cladistic social analysis, 67
Cladogram, 67
Collaboration, 19
Collective identity, 55, 57
Complexity, 25, 31
Complex systems, 109, 129
Computational modelling, 1
　archaeology, 2
　disciplines, 6, 7
　discursive and mathematical models, 18
　dynamic interdisciplinary dialogue, 8
　dynamics, 16
　evolutionary thinking, 9
　generate knowledge, 19
　qualitative variables, 4
　nature/society interactions, 16
　simulation, 19
Computational system, kinship relations, 138, 140
Computer models, 18, 45
Computer simulation, 68
Constraints, 19, 41
Contradict theories, 20
Craft techniques, 168
Cultural consensus, 59, 68–72

Cultural-evolutionary processes, 92
Cultural diversity, 62
Cultural DNA, 61
Cultural homogenization, 63
Cultural identity, 58
Cultural proximity, measurement, 72–75
Cultural relatedness, 67
Cultural transmission, 93
Culture
  definition, 58
  measurement
    conceptual spaces, 61
    distribution of social activities, 59
    ethnic identity, 61
    identity, 58
    in archaeological record, 63–68
    memes, 60
    n-dimensionally variable, 60
    polythetical entity, 62
    Tversky's approach, 60
    variance, 59

D
Dairying
  cattle-based dairying economies, 96
  coevolution of Neolithic, 98
  digest lactose, 102
  Neolithic, 99
  product consumption, 98
Data and variable efficiency, PRECIUM, 29
Deep history of Kinship
  data, hunter-gatherer societies, 137
  evolutionary events, 137
  family space
    behavior patterns, 150
    father relation, 153, 154
    formation of new relation, 151, 152
    mother relation, 150, 151
    recursion, individual to group level traits, 152, 153
    spouse relation, 154, 155
    transformations, 149
  formation, cognitive constraint
    chimpanzee communities, 145
    reduction, social complexity, 145, 146
    social brain hypothesis, 145
    social organization, 145
  genealogical relations (*see* Genealogical relations, kinship system)
  Homo sapiens, Africa, 138
  in human societies, 138
  ontology (*see* Ontology of Kinship relations)

primate social complexity, individuated behaviors, 144, 145
primate social systems, 137
relation-based social systems, 137
STWM (*see* Short term working memory (STWM))
transformation, 138
upper paleolithic, 138
Diet, 94, 96, 102
Discursive models, 18

E
Ecological inheritance, 94
Ecosystem engineering, 94
Ehnoarchaeology
  19th-century intensive farming, 211
  archaeological simulation, 212
  constraining factors, 185
  human environment interaction (*see* Human environment interaction in uplands)
  human impact, mountain ecosystem, 185
  implications, 211
  land-use change (*see* Land-use change)
  premodern practices and ecological knowledge, 186
  simulation outcomes, 208
  upland landscapes, Val di Sole (*see* Italian Alps)
  validation and calibration, 212
  vegetative period, 185
Eigenvalues, 65
Environmental constraints, 43
Environmental modification, 98
Epistemological challenges, 5
Epistemological formalism, 22
Epistemological reflexion, 7
Epistemology, 5
Ethnic groups, 61
Ethnic identity, 57, 59, 61, 66, 67
Ethnicity
  archaeology, 55
  culture, 58
  historical process, 57
  history of interactions, 56
  ideological and political constraints, 56
  indissoluble historical component, 57
  in-group homogeneity, 56
  measurement in archaeological record, 63–68
  social agents, 55, 56
  social identity, 55
  vision, 58
  within-groups heterogeneity, 56
Ethnogenesis, 57, 80

Index

F
Firewood harvesting, 42
Forest foragers (FF), 121, 124

G
Genealogical relations, kinship system
    complexity, 156
    difficulty, tracing, 156
    formation, 155
    Kin Term Space
        determination, 157
        forming, 157
        upper paleolithic, 157–159
    kinship terminologies, 157
    limitations, 156
Ghassulian sites with technological traditions
    local networks (*see* Local networks, Ghassulian)
    shared ceremonial objects
        agropastoral communities, 170
        different scales, 173
        macro-regional distribution, 172, 173
        regional distribution, 172
        size of, 170
        social relationships, 170
        symbolic norms, 171
Go-and-return between fields and models, 221
Granger causality, 99, 102
Granularity, 17
Group alliances, 93

H
Hereditary social inequality, 95
Hereditary socioeconomic inequality, 94
Holistic aspiration of archaeology, 218
Hominin evolution
    face-to-face interaction, 159
    father relation, 138
    "grandmother" hypothesis, 151
    Homo sapiens, 138
    stone artifacts, 148
    STWM, 147, 149
Human environment interaction in uplands
    direct/indirect intervention, 186
    ecological and behavioural constraints, 187
    late prehistoric period, 186
    narratives, 187
    premodern high-altitude practices, 187
    seasonal farming (*see* Seasonal farming)
    vegetative period, 186
HUman Migration and Environment (HU.M.E.) model
    agent-based framework, 119
    Bantu progression, 118
    behavioral rules, 119
    calibration of the spatial quantities, 129
    calibration process, 130
    differentiation, FF and Bantu groups, 131
    diverse populations and environments, 121
    "elastic" metaphor, 131
    energy, 120
    entities, properties, relations and processes, 118
    evolution of behavior, 131
    exchange of food resources, 118
    experimental protocol, 126
    FF and Bantu groups
        differentiation, 130
        direct interactions, 122, 123, 126
    forest foragers (FF), 121
    formalization, 128, 130
    genetic and linguistic literature, 128
    interactions, 119
    knowledge of linguists, 129
    mechanisms, 119
    modelers and thematic experts, 120, 128
    moving without reason, 120
    peopling of an unoccupied land, 118
    peopling process, 120
    resource exploitation, Bantu and FF groups, 121, 123
    rules, 127, 128
    simulation, 126
    spatial and temporal scale, 128
    specification, 127
    validity, 131
Hunter-gatherers, 68, 78, 80, 81, 96
Hypotheses, 32

I
Inductive, 18
Inherited homologies, 66
Institutional evaluation, 220
Interactions, forest foragers, 109–111, 113, 117, 121
Interdisciplinary approach
    accuracy *vs.* precision, 22
    disciplines, 21, 23
    epistemological formalism, 22
    "loud and clear" formalized, 21
    plausibility, 22
    reproduction of domination, 23
    socio-anthropological rules of inheritance, 21
    systemic approach, 21

Interdisciplinary community, 219
Interdisciplinary-formalized archaeology modelling
  qualitative assessments, 219
  simulations, 219
  validation of models, 219
Internal change rate (IRC), 72
Italian Alps
  18th century, 194
  external factors, 193
  population growth, early-modern period, 193
  pre-industrial pastoral and farming practices, 194, 195
  pressure on natural resources, 193
  rural landscapes, 190
  socio-economical transformations, 193
  Val Molinac and Val Poré, 192
  villages of Ortisé and Menas, 190

K
Keep It Descriptive, Stupid! (KIDS) models, 17
Kinship Algebraic Expert System (KAES), 9
Kinship terminologies
  Australian, 138
  form of, 138
  size, 157
KISS models, 6
Kiss *vs.* Kids, 3
Kübler-Ross five stages, 220

L
Lactase persistence, 97, 99
Lactose tolerance, 96–98
Land-use change
  adaptation strategies, small-scale societies, 187
  alpine economy, 210
  assessment, 210
  computational models, 188
  computer modelling, 189
  high-mountain environments, 211
  human behaviour, 187
  Italian Alps, 189
  local economy, 211
  mid-altitude meadows, 210
  and patterns of landscape, 188
  positive surprise of, 210
  premodern farming practices, 211
  quantitative ecology, 188
  reconstructions, 189
  scenario, 210
  simulation scenario, 189

and upland economic transformations
  calculations, 203
  crop productivity, 205
  dairy production, 204
  energy contributions, 205
  haymaking and meadow areas, 206
  high-altitude grazing land area, 205
  labour costs, 206, 207
  low-altitude fields, 206
  non-transhumance species, 204
  parameters, 197–202
  simulations, 202
  static mathematical models, 196
  transhumance species, 204
  wild animals, 204
  wool and dairy production, 207, 208
Late Chalcolithic societies, southern Levant
  analytical sociology, 165
  computational tools, 165
  cultural groups/phylogenetic links, 165
  Ghassulian communities, 166
  Ghassulian culture (*see* Ghassulian sites with technological traditions)
  hypothesis, 166
  interpretation, evolution traits, 164
  microevolutionary processes, 163
  modeling evolution processes
    phylogenetic (*see* Phylogenetic models and technology)
    sociological models (*see* Reference sociological model)
  modes of transmission, 163
  morphocentric, 164
  network structures, 164
  regularities, 165
  relational structure of societies, 164
  rural communities, 165
  technological traditions
    (*see* Technological traditions, late Chalcolithic societies)
Literary models, 2
Local networks, Ghassulian
  ceramic assemblages, 174
  chaînes opératoires
    connecting sites, 174
    population structure, 175
  composition, ceramic assemblages, 175
Lotka-Volterra dynamics, 19

M
Mathematical modeling, 18
Migrations and land use, 30

Index

Modeling methodology
  avoiding initial distortion, 35
  collaboration, 19
  complex rules, 19
  confidence-building test, 37
  experimental approach, 37
  hypotheses and question, 16
  KIDS, 17
  margin of error, 38
  phases, 37, 38
  pyramid of dependencies, 36
  questions, building, 20
  RSES (*see* Rural socio-ecological systems (RSES))
  variability, 38
Models
  constraints, 29
  hypotheses, 17, 38
  KIDS, 17
  rural Neolithic socio-ecological systems, 17
Morisita-Horn similarity index, 65
Multi-agent models, 2
Multi-agent systems (MAS)
  Bantu expansion, 109
  Bantu farmers, 109
  Bantu migrations (*see* Bantu migrations)
  co-construction, 132
  HU.M.E. model (*see* HUman Migration and Environment (HU.M.E.) model)
  hypothesis, 118
  simulation models, 117

N

Neolithic
  agriculture, 102
  analysis, 99
  ancestors–dairy products, 102
  Anthropocene, 92
  burial practices, 92
  cattle ownership, 97
  coevolution, 98
  cultural legacies, 91
  dairy farming and cereal cultivation, 96
  diet and lifestyle, 96
  environmental change, 92
  farmers, 96
  feedback loops, 102
  fertility rates, 92
  human skeletons, 94
  livestock wealth, 92, 97
  niche construction (*see* Niche construction)
  population, 92, 100, 101
  skeletal assemblages, 100
  socio-ecological systems, 17
  village populations, 92
Network topology
  examination, embeddedness, 169
  heterogeneous assemblages, 169
  homogeneous assemblages, 169
  macro-regional scale, 170
  measures of similarity, 169
Niche construction
  adaptive symmetry, natural selection, 93
  capacity of organisms, 93
  domestic animals and crops, 95
  dramatic shift in diet, 94
  ecosystem engineering, 94
  faunal evidence, 94
  habitats and resources, 93
  human diet and land management, 94
  isotopic evidence, Neolithic skeletons, 94
  lactose tolerance, evolution of, 96–98
  loess soils for farming, 95
  non-loess isotopic signatures, 94
  path diagram, 97
  social factors, 93
  strontium isotope
    measurements, 94
    signatures, 95
  testing
    analysis of Neolithic dairying, 99
    decay in probability, site discovery, 100
    density of domestic animals, 101
    evolution, 98
    Granger cause, 99
    lactase persistence, 99
    linear-dynamic-system assumption of equations, 100
    linear modelling approach, 98
    Neolithic skeletal assemblages, 100
    optimal foraging model, 101
    parameters, matrix, 100
    population's capacity, 99
    pragmatic judgment, 98
    radiocarbon dates, 100, 101
    vector-matrix form, 100
  transforming selective environments, 94

O

Ontology of Kinship relations
  categorization, 142
  culture bearers, 142
  elements, 143
  English kinship terminology structure, 142
  external factors, 139
  family space structure, 143

Ontology of Kinship relations (*cont.*)
  human societies, 139
  kin term, 140–142
  paleontological evidence, 139
  parent-child relations, 139
  reciprocal, 143
  structure, 142
  terminologies, 140
Optimal foraging model, 101

P
Parochialism, 56
Peculiarities and traditional methodologies, 15
Phylogenetic models and technology
  chaînes opératoires, 176
  cultural lineages, 176
  measures, 177
  Neolithic and Ghassulian ceramic
      assemblages, 177
  qualitative approach, 176
  southern Levant, Ghassulian period, 176
  variants, 177
Pointillism *vs.* impressionism, 3
Polarization, 77
Population structure, southern Levant, 175
Post hoc ergo propter hoc, 3
Preliminar results, hunter-gatherer, 78, 79, 81
Primum non nocere principle, 31
Processual archaeology, 217
Progressive cultural homogenization, 62
Progressive model, 20
Proximity, 33

Q
Qualitative approach, 27
Qualitative factors in modelling
  failures and struggles, 221
  KISS and KIDS, RSES, 222
Qualitative transformation, Kinship, 137, 149, 155
Quantitative data, 3, 31

R
Radiocarbon dates, 100, 101
Reference sociological model
  archaeological data, 178
  hierarchical formation and political
      power, 178
  interpretation, 178
  networks, 178
  populations, 178
  technological analysis, 178
  variables and interactions, 179
Research community, 45
Research question, OSQHYT
  hypothesis, 26
  object, 26
  question, 26
  subject, 26
  test, 26
Resilience, 66
Roux's epistemological and methodological
    reflection, 10
Rural societies, 19
Rural socio-ecological systems (RSES)
  archaeological and paleoenvironmental
      knowledge, 16
  computational models, 16, 18, 19
  construction of model, difficulties, 20
  distributive modelling
    agent-based models, spatial
        reconstruction, 23, 24
    equilibrium, simplicity and complexity,
        24, 25
    interdisciplinary approach (*see*
        Interdisciplinary approach)
  epistemological benefits and constraints, 18
  hypotheses question, 16
  migrations and land use, 30
  social components, 30, 32, 38
  variables, 28

S
Seasonal farming
  activities, 186
  altitudes, 202
  crop productivity, 205
  dairy production, 203
  low-altitude fields, 206
  non-transhumance species, 204
  transhumance species, 204
  yields, areas and energy contributions, 205
Sensitivity/robustness
  efficient and discriminative
      sensitivity, 29
  perennial data sources, 29
  robustness of measurements, 29
Short term working memory (STWM)
  changes, cognitive abilities, 147
  conchoidal flaking, 148
  qualitative design change, 149
  size of, 146
  stone artifacts, 148
  technology of artifact production, 149

Index

Simulation
  agents, 73
  approaches, 218
  Bantu expansion (*see* Bantu expansion)
  computer, 68
  cycle, 72
  frozen state, 63
  models, 24, 27
  neutral identity, 72
  output, 72
  statistical indexes, 72
  virtual groups, 81
Social agents, 56, 58, 59, 61, 66, 67
Social and communication usefulness
  expressive, 28
  generic, 28
  inter-comprehensive, 29
Social animals in hierarchical community, 23
Social brain hypothesis, 145
Social component affecting binaries
  enlarged families/mononuclear families, 41
  inheritance and gender stratification, 40, 41
  practices and geography, 41
  ultimogeniture *vs.* primogeniture, 41
Social conflicts, 78
Social constrains, 56
Social diversity, 78, 82
Social fractionalization, 75–78
Social identity, 55
Social inertia, 74
  aggregate of social agents, 66
  cladogram, 75
  cultural relatedness, 67
  empirical measures, 72
  global measure, 72–75
  temporal evolution, 74
Social interaction, 10
Social modelling
  binaries (*see* Social component affecting binaries)
  economic activities
    agriculture, 42, 43
    fishing, 45
    hunting, 44, 45
    livestock-keeping, 43, 44
    wood harvesting/cutting, 42
  scale gap, 39–40

Social reproduction mechanisms, 56
Social tension, 77
Socio-anthropological field methodologies, 5
Socio-anthropological rules, 21
Socio-ecological modeling
  social component
    hierarchy criteria, 32–35
    rationality and structure, 30–32
Socioecological modelling, 4
  landscape and land-use change, 188
  pre-industrial rural strategies, 188
  small-scale societies, 187
Socioecological models, 5
Socioecological systems, 222
Socio-environmental psychology, 16
Socio-natural models, 19
Spatial hypotheses, 42
Strontium isotope signatures, 95
Sustainability
  and resilience, 187, 188
  ecological, 189, 211, 212
  premodern high-altitude practices, 187

T

Technological traditions, late Chalcolithic societies
  exchange-based relationships, 167
  measure network topology
    embeddedness, 169
    techno-petrographic analysis, ceramic assemblages, 169, 170
  network analysis, 166
  social interactions, 167
  variables, socially link individuals/groups
    anthropology of techniques, 168
    ceramic *chaîne opératoire*, 167
    community of practice, 168
    cultural transmission, 168
    forming techniques, 168
    spatial patterns, 168
Tversky's approach, 60

U

Utility threshold, 69